SOLAR SYSTEM

SOLAR SYSTEM
Between Fire and Ice

Thomas Hockey, Jennifer Lynn Bartlett, and
Daniel C. Boice

CRC Press
Taylor & Francis Group
Boca Raton London New York

CRC Press is an imprint of the
Taylor & Francis Group, an **informa** business

CRC Press, Boca Raton and London
by CRC Press
6000 Broken Sound Parkway NW, Suite 300, Boca Raton, FL 33487-2742

and by CRC Press
2 Park Square, Milton Park, Abingdon, Oxon, OX14 4RN

Library of Congress Cataloging-in-Publication Data
Names: Hockey, Thomas A., author.
Title: Solar system : between fire and ice/Thomas Hockey,
Jennifer Lynn Bartlett, and Daniel C. Boice.
Description: First edition. | Boca Raton : CRC Press, 2021. |
Includes bibliographical references and index.
Identifiers: LCCN 2021023389 | ISBN 9780367768690 (paperback) |
ISBN 9781032054377 (hardback) | ISBN 9781003197553 (ebook)
Subjects: LCSH: Outer space–Exploration. | Solar system.
Classification: LCC QB500.262 .H63 2021 | DDC 523.2–dc23
LC record available at https://lccn.loc.gov/2021023389

ISBN: 978-1-032-05437-7 (hbk)
ISBN: 978-0-367-76869-0 (pbk)
ISBN: 978-1-003-19755-3 (ebk)

Typeset in Times
by MPS Limited, Dehradun

Front cover credit - Steve Gildea, *Planetary Suite
(detail)*, 1992, Oil on canvas panels, Collection
of Merrimack College. Reproduced courtesy the artist,
Suite3D.com.

Back cover credit - : NASA/JPL-Caltech/MSSS

Dedication

T. H. семье Зиминых, которая приняла меня в свой клан.

J. L. B. For John Howard Riggleman, as grandmother and grandfather told me, remember the sky is not your limit.

D. C. B. My fortune cookie said, "A Partner Can Help You Achieve Success." To Panida Boonmasai for your constant encouragement and support. รักคุณมาก.

Table of Contents

Preface

Our understanding of the Solar System has changed in only a few short years. The authors of this book, the first of its kind for the 2020s, strive to make it the most up-to-date popular resource in the field. We attempt to do so with lively, conversational prose and bountiful illustrations. It took three of us to accomplish our task!

Solar System: Between Fire and Ice takes a unique approach to our neighborhood in space. We do not follow a strictly in-out trajectory from the Sun. Instead:

- We begin with the ultimate beginning, the formation of all energy and matter, and show more than one way in which our Solar System might have come into being.
- After a description of the Solar System as a whole, we are outward bound to the planets Mercury and Venus, which today merit their single-subject chapters.
- However, we start, not with the first planet, but instead with the Moon: the most well-known world beyond the Earth and one that we intend to revisit soon.
- Asteroids and meteors also get their due, especially the growing concern about catastrophic impacts and what to do about them.
- Some books give short shrift to the outer Solar System. Not this one. Here is where some of the most exciting discoveries of the 21st century are being made.
- For instance, the satellites of the outer planets, full-scale worlds themselves, get the separate treatment they deserve.
- The popular subject of comets follows, acknowledging the relationship between all these frozen bodies and the important clues they hold about our origins.
- We consider the dwarf planet Pluto from the point of view of just one of many icy worlds outside the orbits of the eight planets.
- Not to be ignored, a complete chapter is devoted to the Sun, the elephant in our Solar System room.
- Planets orbiting stars other than the Sun, now known to be plentiful, deserve their own chapter.
- We end with a one-of-a-kind timely consideration of humankind's possible future in the Solar System.

Along the way, we include the most up-to-date results from new telescopes and spacecraft. At the same time, we do not neglect the historical perspective and our cultural connection to the planets. It will be a wonderful voyage.

Hang on!

Thomas Hockey, Jennifer Lynn Bartlett, Daniel C. Boice—May 2021

Foreword

Dr. Reta F. Beebe
Emeritus University Professor of Astronomy, New Mexico State University
Recipient of NASA's Exceptional Public Service Medal

In the 1980s and 1990s, the Viking and Voyager missions provided droves of exciting information, generating a new level of public interest. Textbooks were rewritten and scientists worked to understand the data during the mission-poor period that followed. In recent times, however, we have entered a new era. There has been a multinational effort to expand our knowledge of the Solar System. Data from these missions has been freely shared and has again raised the level of public interest. Within this era of renewed interest, it is appropriate, as is done in this book, to provide the public with an effort to present an integrated view of our Solar System and questions that the discovery of extrasolar planets have raised with regard to the Solar System as a whole. While Martian rovers and orbiters with greatly improved instrumentation provide us with a more refined understanding of Mars, recent missions to the outer planets have stretched our imagination when we try to understand how physical processes are manifested by changing solar distance and varying conditions of formation of the planets.

Results from the Saturn Cassini mission have given us insight into how a complex planet-rings-satellite system has evolved and functions and presented Titan to us as a strange and intriguing sister of our earth. The New Horizon mission showed us that Pluto is unexpectedly complex and introduced us to cold volcanism, while the Juno mission has greatly expanded our knowledge of Jupiter, destroying our picture of the interior, providing us a better basis for understanding planetary magnetic fields and revealing a view of a complex polar weather system.

It is within this context that the authors present you with a summary that addresses how the planets have been shaped by variations of shared properties and location in the Solar System. This will provide an integrated basis for understanding trends and peculiarities in the Solar System but may leave you wanting more about individual bodies. So, while reading this book or after completing it, if you feel the urge to delve deeper into information concerning a particular body, you may worry about the validity of what you could find on the web. A direct way to start is to search the web by entering NASA (for US), ESA (for Europe), JAXA (for Japan), ISRO (for India), or UAE (for United Arab Emirates) and the name of the body or mission that you seek. You'll encounter pages set up for you and be able to branch out enough to be your own critic. Welcome to my worlds and happy adventures!

Acknowledgments

We wish to thank the following scientists for reading a pre-publication draft of this work: Alan Hirshfeld, Amaury de Almeida, Catherine Garmany, Daniel Green, Jonathan Keohane, Kevin Marvel, Reta Beebe, Nancy Chanover, Steven Maran, Steven Bloom, and Randy Gladstone. Errors are the responsibility of the authors.

This book further benefitted from my discussions with Amy Simon, Clark Chapman, Dale Cruikshank, Donald Yeomans, Eileen Ryan, James Lattis, Jay Pasachoff, Kenneth Rumstay, Kevin Schindler, Patrick Seitzer, Sylvia Baggett, Timothy Cooney, William Kaufman✝, and William Sheehan. I appreciate assistance from Yuliana Ivakh, Siobahn Morgan, Noel Graff, and Sydney McFee.—T. H.

The comments of Daniel Bliss, Cathy Walker, Kristy Johnson, Barbara Natalizio, and Gabrielle-Ann Torre helped me find a voice and structure for this project and without their encouragement this book would be never taken the shape it has. The training and mentoring provided Phil Ianna, Mark Whittle, Ed Murphy, Charlie Tolbert, and Bob Rood✝ at the University of Virginia forever changed my life and enabled me to pursue astronomy professionally. The professional example of Alice Monet continues to inspire me. As important as each of these influences are in shaping my career and this book, none of it would be possible without the immeasurable love and support of my parents, Jane and John Bartlett, my husband, Kenny Riggleman, and my son, John Howard Riggleman. Even more fundamentally, I am grateful to my Creator for each fascinating world in our amazing Solar System and the millions to be explored beyond.—J. L. B.

I am grateful to my thesis advisor, Professor Herb Beebe (New Mexico State University), who guided me along the arduous (but enjoyable!) passage through graduate school. The collegial relationship with my mentor, Dr. Walter Huebner (Southwest Research Institute), and his valued friendship during the past four decades, is deeply appreciated. I benefitted greatly from the love and support of my wife, Panida Boonmasai, my daughter, Michelle Garza, and her husband, Luis Garza. Acknowledgments must include the sponsorship of NASA and the National Science Foundation who supported my career, and the opportunities provided by my employer of 26 years, Southwest Research Institute. Ultimately, I must thank the American people for the trust that their money would be well spent.—D. C. B.

Author Biographies

Thomas Hockey was born in the same year that a robot space probe photographed the heretofore hidden farside of the Moon. He clearly remembers first spotting elusive planet Mercury in the sky and seeing Saturn's rings through a small telescope in his youth. It was a time when voyaging into the Solar System dominated the media. Dr. Hockey recognized that he did not match the physical appearance of a contemporary space adventurer: a tall, square-jawed, and buzz-cut astronaut of the 1960s. However, he had an epiphany when he was told that there was another way to explore other worlds, through the discipline of astronomy.

Dr. Hockey studied Planetary Science at the Massachusetts Institute of Technology. There he worked for Professor James Elliot, discoverer of the rings of Uranus. His doctoral work took place at New Mexico State University, under the mentorship of Professor Reta Beebe, award-winning member of the historic Voyager space-probe Imagining Team. He also counts among those who influenced his career Professor Herbert Beebe and Professor *Emeritus* Clyde Tombaugh, the astronomer who initially spotted the dwarf planet Pluto.

Dr. Hockey joined the faculty of the University of Northern Iowa where he has taught astronomy to ten thousand students. He is the author of numerous professional papers and books. He is most well-known for serving as Editor-in-Chief for the prize-winning *Biographical Encyclopedia of Astronomers* (four volumes, editions 1 & 2, Springer) and for writing *How we See the Sky: A Naked-Eye Tour of Day and Night* (University of Chicago Press). Dr. Hockey also edited the journal *Astronomy Education Review*. He is a Fellow of the Royal Astronomical Society and member of the International Astronomical Union. In 2017, asteroid (25153) 1998 SY53 was named **Tomhockey**.

Dr. Hockey works with its Chair, Dr. Jennifer Lynn Bartlett, on the American Astronomical Society's Working Group for the Preservation of Astronomical Heritage. He was a graduate-school classmate of Dr. Daniel C. Boice at New Mexico State. 'Go Aggies.'

Dr. Hockey is married with two sons and two step-daughters. He has visited nearly 40 countries, including expeditions to observe seven total eclipses of the Sun. His hobbies include collecting planispheric celestial volvelles ('star wheels'). His favorite color is purple, that of the most luminous stars.

Jennifer Lynn Bartlett is an astronomer with the US Naval Observatory, where she computes the positions and motions of planets and other celestial bodies while promoting traditional celestial navigation. She is also currently a member of the Organizing Committee for the International Astronomical Union (IAU) Commission A1 Astrometry. Previously, she taught introductory astronomy at Hampden-Sydney College and had a professional career as an engineer.

Having never lost her childhood love of the night sky, **Dr. Bartlett** left engineering to pursue graduate studies in astronomy at the University of Virginia. As part of her dissertation research under the direction of Philip A. Ianna, she detected nearly one exoplanet, investigated early reports of planets orbiting Barnard's Star, and observed 2MASS J23062928-0502285 (now popularly known as TRAPPIST-1) extensively. Although she continues to investigate cool stars and their companions, her most recent research involves understanding how the terrestrial atmosphere affects observations.

Having developed a passion for the history of astronomy as a graduate student, **Dr. Bartlett** is now chair of the American Astronomical Society Working Group on the Preservation of Astronomical Heritage and an oral history interviewer. In addition, she is a member of the IAU

Commission C3 History of Astronomy. She met fellow author Thomas Hockey through the history of astronomy community. Having convinced her to come on this adventure with him, she is succeeding him as one of two editors-in-chief of the next edition of the *Biographical Encyclopedia of Astronomers.*

When neither stargazing nor thinking about stargazing, Dr. Bartlett enjoys reading, swimming, and archery. She is eternally grateful for her husband, who appreciates seeing Saturn, and her son, who enjoyed seeing Comet NEOWISE.

 **Daniel C. Boice** is the principal astronomer at Scientific Studies & Consulting in San Antonio, Texas. Prior to this position, he performed cometary research sponsored by NASA and the National Science Foundation for 26 years in the Space Science and Engineering Division at Southwest Research Institute. Concurrently, he held a faculty appointment for 20 years in the Department of Physics and Astronomy at the University of Texas at San Antonio, where he taught undergraduate and graduate courses to our next generation of astronomers.

Following his childhood passion for the night sky, **Dr. Boice** received his BS in Physics at Brigham Young University in 1975. He obtained a PhD in astronomy at New Mexico State University in 1985, where he met co-author, Dr. Thomas Hockey. While a postdoctoral fellow in the Theoretical Division at Los Alamos National Laboratory and at the Max-Planck-Institut für Astrophysik in Garching, Germany, Dr. Boice developed computer models that have been successfully used to interpret spacecraft data and observations of many comets. He was a member of the science team for NASA's Deep Space 1 Mission to Comet 19P/Borrelly.

Dr. Boice's professional activities include over 80 peer-reviewed research papers and several hundred conference reports. He has served as Past Chairs of the Physical Studies of Comets Working Group (International Astronomical Union) and Space Related Studies of Small Bodies of the Solar System (Committee on Space Research). An emeritus member of the American Astronomical Society, in 2000, he became a Fellow of the Royal Astronomical Society. He has spent several years abroad teaching and working with colleagues in Germany, Japan, France, and Brazil. In recognition of his work in cometary science, asteroid 7195 **Danboice** was named in his honor.

Dr. Boice continues his comet research to assimilate information from ground-based observations and *in situ* spacecraft measurements to develop a better global understanding of comets. With co-author Hockey, *Comets in the 21st Century: A Personal Guide to Experiencing the Next Great Comet!* (Morgan & Claypool) was published in 2019. He is married to Panida Boonmasai with two daughters, Michelle and Stephanie, a son, Christopher, a step-son, Tam, and a granddaughter, Alison. When not engaged in all things comets, Daniel loves collecting books and rock 'n' roll music, board gaming, and rice farming with his family in Northern Thailand.

Introduction

In this book, join us as we tour the Solar System, stopping as needed to look at how scientists arrive at astronomical knowledge. We point out new discoveries and understandings along with the things that remain very much a mystery to us. All this is to say that, far from doing something specific to one narrow scientific discipline, we do what each of us thinking human beings has done since the beginning: try to figure out where we are and what the 'rules' are. In this regard, planetary science is a very human endeavor.

We present a number of ideas that we hope you still will consider even when the specifics of *Solar System: Between Fire and Ice* begin to fade. Whether explicitly or implicitly, we return to these ideas again and again throughout the book:

- Sorting natural objects and phenomena into categories, classifying them, is a useful first step at explaining and interpreting them. We must be careful, though, to distinguish between apparent properties and intrinsic ones.
- The story of many natural systems is one of gradual change punctuated by occasional, sudden, course-altering events. Both equilibrium and cycles are common in the Solar System.
- Objects, such as planets, are not unchanging; they evolve with time. Their appearances today offer clues as to their histories.
- Our lives and environment here on the Earth bias us as to what a typical place in the Solar System is like.
- Our ability to affect the Solar System is minuscule. The Earth will continue with or without us. Humans have altered the Earth's climate since the industrial age, causing increased global temperatures on average. Maintaining the unusual set of conditions that allows for life on the Earth to prosper is in our own self-interest. No one will 'bail us out' if we fail, and we have no place else on the seeable horizon to go.

Our tour follows the planets outward from the Sun beginning with Mercury after a detour past the Moon. However, like astronauts preparing for a mission, we first spend some time, just three chapters, learning the tools we will need. Once we have stopped at each of the eight major worlds and some of the minor ones as well, we return to the center of it all, our Sun. Because our Solar System is a small fraction of the worlds to be discovered, and because planetary science still has much to learn within its bounds, we find ourselves concluding this book where the future begins.

Origins. Where do we come from? The desire for an answer to this question is universal. Different peoples have responded in distinct ways. But to understand our origin in the modern sense of the question, we must try to understand that of our home, the Solar System.

Planets. Certain truths apply to all the planets. Such basic quantities can be bootstrapped to learn more about these worlds. So, what at first glance might seem like, a sea of numbers actually informs us about the personality of a planet. We unpack the numbers and with them paint a picture of what to look for in chapters ahead.

Orbits. The Sun holds its retinue of worlds closely to its heart; it is this grip that defines our Solar System. Without the Sun, planets would be doomed to wander in the dark cold depths of space. A planet's place in the Solar System greatly affects its nature. Gravitation is the glue that binds these bodies to the Sun, ever traveling in bound paths called orbits.

Moon. The first stop beyond the Earth is the Moon. In the aftermath of the Apollo program, the Earth's own satellite seemed neglected or forgotten. However, lunar science and exploration are having a renaissance with as many as ten missions sent to the Moon during the past decade.

Among these are landers operating on the hitherto unvisited hemisphere facing away from the Earth.

Mercury. Long an enigma because of its proximity to the Sun, Mercury is hard to view from the Earth. Nevertheless, robotic space probes are revolutionizing our knowledge of this comparatively near but seldom-visited, innermost world. Astronomers now have tentative knowledge of the planet's internal structure, surface features, and harsh, space environment near the Sun.

Venus. Is Venus really the Earth's twin? Based on current and past observations from the Earth and the planet itself, we present everything you need to consider in your decision: its interior, surface, and atmosphere.

Earth. How to treat the Earth as just another planet? Our very definition of what a planet ought to be like is skewed by the one upon which we happen to live. Here, we point out how the Earth differs from other planets in important ways, conditions that make it so well-tailored for the development and evolution of life.

Mars. Because it is the most Earth-like of planets, Mars continues to be a compelling planetary exploration target. As of this writing, the martian neighborhood hosts 8 operating orbiters, 2 landers, 2 rovers and a helicopter. These space probes explore the martian atmosphere, surface, and subsurface to understand the planet's history and potential habitability.

Asteroids and Meteoroids. These small bodies are rocky/metallic leftovers from the formation of the Solar System. Two space missions recently brought miniscule samples of interesting asteroids back to the Earth; one more mission is attempting to return material from a third asteroid. Although Near-Earth Objects present a serious threat to us with their potential violent impacts, they also are exciting possible resources to extend our mineral reserves.

Jupiter and Saturn. Two worlds are truly masters of the Solar System: Jupiter and Saturn. They are the biggest. They are the most massive. They are the first of the giant planets. Epic, robotic, space voyages, requiring years of travel, inform us about these Gas Giants, plus their associated rings, satellites, and extreme magnetic fields. Juno to Jupiter and Cassini to Saturn are the most recent fabulous journeys of outer-solar-system exploration.

Uranus and Neptune. These Ice Giants are among the least explored worlds in our Solar System, and, yet, they may represent a planetary class that is commonplace in the Milky Way Galaxy. Despite many similarities, astronomers observe significant differences between the two: a riddle at the edge of the planetary family.

Icy Satellites. We humans now have the opportunity to search for signs of life in one or more of the ice-covered ocean worlds of the outer Solar System. These frozen bodies are not planets but instead satellites orbiting the giant planets. They include exotic Titan with its lakes, streams of liquid hydrocarbons, and solid organic molecules, from which living organisms are formed on the Earth.

Comets. The 'delinquents' of the Solar System, the comets, swoop down upon the planets from the far reaches of the Kuiper Belt and beyond. These minor icy bodies nonetheless produce magnificent tails, making them iconic features in our sky. Comets are the most primitive bodies in the Solar System and, as such, hold clues to its origin and possibly to the genesis of life itself.

Kuiper Belt Objects and the Oort Cloud. Once upon a time, just Pluto was known to revolve at the edge of the Solar System. Astronomers now understand that the realm beyond the outermost planet is inhabited by small, icy bodies that represent the Solar System in its most primitive state. Moreover, the Kuiper Belt and the much farther Oort Cloud are home to those interlopers in the Planetary System: comets.

Sun. Our Solar System takes its name from Sol, an archaic word for the Sun, its defining body. A source of light and heat for the planets that orbit it, and are dwarfed by it, the Sun produces energy in a way that we only can dream about so powering human endeavors. Yet, the Sun is simply a close-up example of the myriad of other stars that populate our Galaxy.

Exoplanets. While not part of our Solar System by definition, we must consider those planets orbiting stars other than the Sun in order to understand fully what a planet can be. Ground- and space-based facilities are revealing a remarkable variety in the observed properties of exoplanets and the planetary systems to which they belong; these different systems express even greater diversity than astronomers already find in our own.

Future. A new space missions soon will further explore Mercury. No less than five target Venus. Planetary scientists are near to collecting samples of Mars for return to the Earth. Farther out in the Solar System, missions will revisit Jupiter to explore the Galilean Satellites. With all this robotic travel about our local Planetary System, can humans be far behind?

Investigating distant worlds is expensive. Time-consuming. Risky. Nonetheless, exploration is an innate urge within us, as humans. And doing so provides a cautionary tale: We inadvertently are changing our planet, rapidly making it less hospitable than the globe we have come to know; instead, we are turning it into a more challenging, alien place. By studying the stories of our solar-system neighbors, we perhaps avoid such a fate.

FIGURE In July 2020, Comet NEOWISE delightfully surprised sky watchers in the Earth's Northern Hemisphere, who had waited more than two decades hoping for a bright comet to appear.

1 Origin of Our Solar System

According to Icelandic poet Snorri Sturluson ⟨1179–1241⟩, the Universe began as a gaping void, the Ginnungagap, suspended between a burning region of flames, Muspell, and a frozen region, Niflheim, that contained a spring from which 11 rivers flowed. The mixing of burning winds from Muspell with rime from Niflheim produced the giant, Ymir, and the cow, Audhumla. A race of frost giants emerged from Ymir, while Audhumla licked a man, Buri, out of the ice. Odin, Vili, and Ve, Buri's grandsons, slew Ymir and created Midgard from his dismembered body. They captured sparks from Muspell to be stars and planets. They carved the first man and woman, Ask (ash tree) and Embla (elm tree), from logs. Odin and his brothers placed the Sun and Moon in chariots to circle the Earth. However, two great wolves chase the Sun and Moon and will swallow them at Ragnarok, the cataclysmic battle that almost destroys the nine worlds.

According to Genesis, the Universe began as formless darkness over which God spoke the world into existence. He commanded light on the first day, the sky on the second day, land with plants and seas on the third day, the celestial bodies on the fourth day, animals of the sea and sky on the fifth day, and land animals and humans on the sixth day. He formed the first man, Adam, from dust and the first woman, Eve, from Adam. On the seventh day, God rested. When Eve and Adam disobeyed one of His rules, God expelled them from the Garden of Eden, in which they were dwelling, into the harsher, but familiar, world.

Our ancestors used stories like these to make sense of the surroundings in which they found themselves and bring order to their lives. American planetary scientist Carl Sagan ⟨1934–1996⟩ described science as, "at least in part, informed worship." Cosmology, the study of the Universe itself, also tries to make sense of our observations and describe the natural laws that function throughout it. One of those laws is that light travels at a finite and measurable speed: 300,000 kilometer per second (190,000 miles per second).[1]

A consequence of this universal speed limit is that we do not observe the world as it is, but as it was. The light reaching your eyes from the hand you are waving in front of your face reflected off your appendage about a nanosecond ago. Your hand probably did not change much during that interval. The light arriving at the Earth from the Sun has been traveling for just over 8 minutes, during which time a flare might have occurred. Similarly, we see the α Centauri[2] system, the next closest system of three stars, as they were slightly more than 4 years ago, so their star spots will have evolved. We see the Andromeda[3] Galaxy, just visible to the naked eye, as it was about 2.5 million years ago. In which time, some of its stars will have died and new ones begun forming.

As we look farther out into the Universe, we see further back in time. Looking outwards in the 1920s, American astronomer Edwin Hubble[4] ⟨1889–1953⟩ reviewed the motions and distances of 46 different galaxies.

American astronomer Vesto Slipher ⟨1875–1969⟩ had determined the motions of galaxies by spreading their light with a prism into their component colors, or their spectra. The study of spectra, or spectroscopy, can reveal many properties of the body that emits the light, including the speed and direction in which it is moving.

Using a method of "standard candles," Hubble measured distances to 20 of these galaxies by comparing the brightness of stars within each to the brightness of similar nearby stars. If two stars are sufficiently alike, then the dimmer one is the more distant one, like streetlamps stretching away from you along an avenue. He calibrated his scale using a particular class of star that varies in brightness, the Cepheids.[5] American astronomer Henrietta Leavitt ⟨1868–1921⟩ had previously established a relationship between the luminosities and the characteristic periods for these pulsating stars.

Once Hubble plotted all this information, his graph revealed that each of those galaxies is moving away from us with the more distant galaxies moving faster than nearer ones. These observations matched theoretical work by Belgian priest and cosmologist Georges Lemaître ⟨1894–1966⟩.[6] Although we do not experience it personally in our daily lives, the Universe is expanding.

Running its expansion backwards, the Universe compacts to an incredibly hot, infinitely dense point. Approximately 14 billion years ago, that point began expanding. In the moment of the Big Bang, the Universe—space, time, energy, and matter—was created. Science cannot tell us what is before the before or beyond the beyond with respect to the Big Bang. Similarly, the Buddha ⟨*circa* 5th century BCE⟩ did not speculate on the origins of the world and discouraged his followers from fretting over unanswerable questions (Figure 1.1).

However, the afterglow of the Big Bang is detectable. After about 300,000 years of cooling, the temperature of the Universe was low enough that electrons, protons, and neutrons could combine into atoms of hydrogen, helium, and lithium. The light released during this recombination event is the oldest science can measure. In 1964, American radio astronomers Robert Wilson and Arno Penzias[7] determined that the irritating buzz in a Bell Laboratory horn antenna was not from pigeon poop as they first suspected but light from this earliest period, its waves stretched by the expansion of the Universe. Since then, the COsmic Background Explorer [COBE] and Wilkinson[8] Microwave Anisotropy Probe [WMAP], two satellites launched by the United States [US/USA] National Aeronautics and Space Administration [NASA], mapped this cosmic background radiation [CBR]. Although the CBR is nearly uniform, these maps reveal minute hot and cold spots (Figure 1.2).

Small variations in the density of the primordial gas were the seeds of clouds that eventually formed galaxies and stars. Within a billion years after the Big Bang, both galaxies and their constituent stars were present throughout the Universe (Figure 1.3).

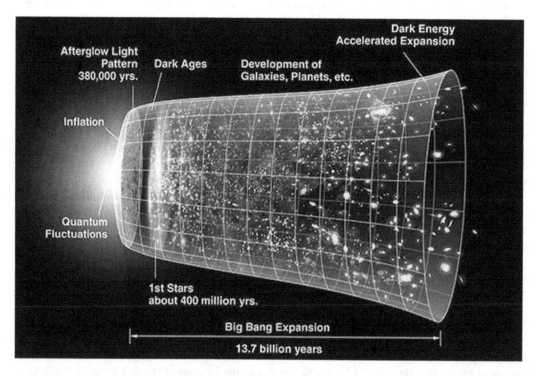

FIGURE 1.1 Representing the expansion of the Universe.

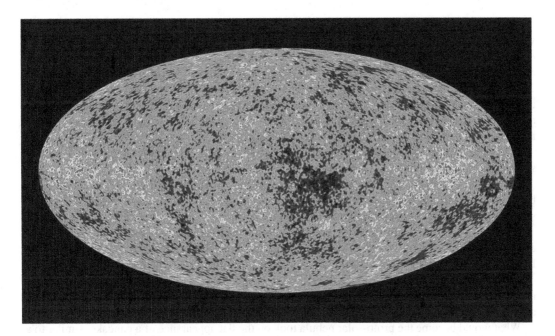

FIGURE 1.2 CBR over the whole sky, if we could see in radio wavelengths.

FIGURE 1.3 Universe of galaxies; each may contain hundreds of billions of stars.

Stars shine because they produce energy by fusing light elements into heavier ones. The most-massive stars live short intense lives before ending in supernovae, massive explosions that drive enriched stellar gas into the surrounding galaxy. The earliest generations of stars included massive ones, which produced most of the chemical elements. Over the next 8 billion years or so, each succeeding generation of stars formed from clouds of gas and dust that included increasing amounts of elements heavier than hydrogen and helium.

About 9.2 billion years after the Big Bang, or a 'mere' 4.6 billion years ago, the Solar System was not yet the complex structure we know today. Instead, it was just beginning as a simple cloud of dust and gas, a nebula, containing hydrogen, helium, and traces of heavier elements. It was one of many amongst the stars of our home, the Milky Way Galaxy. This cloud slowly orbited the center of the Milky Way. As the nebula moved, it carried along its dust and gas, all of which were also moving within it. The net effect of these motions was a slight rotation of the gas and dust around its center. This cloud, called the proto-solar nebula, started to collapse under the mutual attraction of its individual particles, or its own gravity (Figure 1.4).

Why? Maybe a passing star gave it a nudge? Maybe a shockwave from a dying massive star—a supernova—struck the nebula? The death of one star begat another.

Most of the matter (99.8%!) ended up in a compact ball, so dense and hot that nuclear reactions began to take place in the center. Hydrogen fused to form helium and released energy. The object shone brightly and eventually was named the Sun.

As clumps formed within the proto-solar nebula, these future worlds moved with a net speed and direction that was the sum of the motions of the smaller particles from which they grew. Because the small particles were generally moving in the direction that the overall nebula was rotating, the larger clumps followed this pattern as well. The motion in the nebula's direction of rotation predominated. This direction was perpendicular to the direction towards the forming Sun, at the center of the nebula.

What would become the proto-solar nebula took on the flat appearance of a pancake, similar to a pizza maker spinning your pie (Figure 1.5). Only a slight statistical surplus of material rotating about one axis, as opposed to any other, will create a disk-like system. As the disk grew, its gravity dragged any material moving around a different axis into alignment with the majority. Alternately, wayward material was left behind as the proto-solar nebula contracted. The nebula rotated faster and faster, just as an ice skater spins up when he pulls his arms and legs together. This simple theory explains why almost every one of its primary members moves about the Solar System in the same direction, and in a more-or-less flat plane (Figure 1.6).

The release of gravitational energy from the coalescing Sun caused it to become hot, to begin to radiate, and to keep things close to it warm. Meanwhile, the outer region of the proto-solar nebula grew cold, as all things do with time in the absence of a neighboring heat source. Regardless of temperature, what was left of the nebula continued to revolve about the proto-Sun. Progressively,

FIGURE 1.4   Solar System forms from a nebula similar to the Eagle Nebula depicted.

FIGURE 1.5   Pizza dough flattens as it spins.

the primitive ingredients of the proto-solar nebula began to cool, condense, and clump together. The clumps grew into a myriad of small bodies called planetesimals approximately 1 kilometer (0.6 mile) in size. The clumps farthest from the hot proto-Sun formed planetesimals first, gathering with them much of the available material including vast amounts of hydrogen (Figure 1.7).

Elements heavier than helium were comparatively rare in the proto-solar nebula. Still, those elements with higher melting temperatures grew cool enough with time to condense into liquids and solids even near the soon-to-be Sun. Volatile elements—those with low melting temperatures—did not. The solar wind, an outflow of charged particles produced by the nascent Sun, blew the lighter elements away from the inner regions of the proto-solar nebula. It left only a few to be captured by slowly forming planetesimals.

Gases condensed to liquids, and liquids condensed to solids. In the cold outer region of the proto-solar nebula, even volatiles froze to ice. Without interference from the solar wind, which weakens with distance, *lots* of volatiles were available. The widely spaced planetesimals in this region had huge volumes of the remaining nebula from which to 'feed' upon and grew extremely large. The proto-planets that formed in this region would be massive, but also larger and so not very dense.

FIGURE 1.6   Representing the Sun condensing from the proto-solar nebula, which is flattening into a disk.

FIGURE 1.7   This small and oddly-shaped object shows the characteristics of a planetesimal left over from the early Solar System.

Material closer to the proto-Sun took longer to cool and clump. By the time these planetesimals formed, most of the hydrogen and helium was collected in the Sun and outer proto-planets by these bodies' gravity. The planetesimals eventually formed there were fewer and made of rock and metal that condense at high temperatures with, perhaps, a thin veneer of volatile gas, if the planetesimals had enough gravity to hold onto it. The resulting inner proto-planets would be less massive and smaller, but, therefore, denser.

Although most by far of the original nebula eventually ended up in the middle as the embryonic Sun, plenty of material remained and grew into planetesimals massive enough to influence each other. A mad game of gravitational billiards ensued with each affecting the orbit of the others. Crashes inevitably occurred. In some cases, collisions produced more, smaller planetesimals, but in others, the planetesimals merged into fewer, more-massive chunks. Some planetesimals avoided smash-ups and pileups but were expelled to the farthest reaches of the Sun's gravitational influence and beyond.

The planetesimals resulting from mergers had greater gravitational influence and grew further. The most successful planetesimals were now less plentiful but more massive. They began crushing themselves into nearly spherical proto-planets, hundreds of kilometers in size.

The slow accretion of planetary cores may have taken a few million years, during which time the solar wind was removing the lighter volatile elements. Other mechanisms may have allowed the giant outer planets to grow quickly enough to trap the significant amounts of hydrogen and other gases that are their primary components. Perhaps instabilities in the solar nebula allowed some clumps to form earlier or clumps the size of pebbles to fuse together more rapidly (Figure 1.8).

Regardless of all this smashing together and knocking out, the result was eight major bodies and trillions of left-over smaller ones ranging from mere dust to asteroids and comets, even dwarf planets. The chemical make-up of these bodies varies with their distance from the Sun. The changing composition reflects the conditions in the proto-solar nebula at the locations where each formed. Science gives the Sun and its acolytes a name, the Solar System (Figure 1.9).

Once formed, planets continue to evolve. In a process known as differentiation, heavier elements, such as metals, tend to sink to the center while lighter materials, such as molten rock, float toward the surface. Slowly, planets and moons cool. Some of these form solid surfaces, or crusts,

FIGURE 1.8 Representing the planetary system forming through the aggregation of planetesimals.

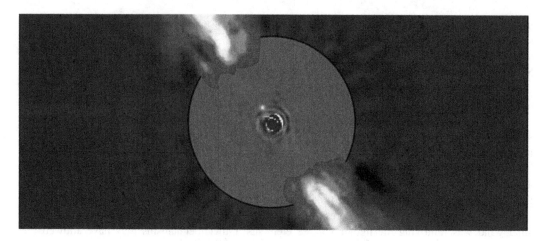

FIGURE 1.9   This infrared image appears to show a system of planets forming about a young star.

of rock or ice, while the most massive seem to transition from a gaseous hydrogen atmosphere to a liquid hydrogen ocean without a clearly defined surface.

The leftover debris is interesting enough to warrant study on its own. ☞ Chapters 9 and 13 ☞ The crusts of planets and moons are scarred by craters resulting from collisions with such rubble. Because each impact is a one-way trip, the rate of cratering has decreased with time. Surfaces throughout the Solar System record an intense period from about 4.5 billion to 3.8 billion years ago, which is known as the Era of Heavy Bombardment. Although the frequency of debris striking the planets is much less now, fresh craters still appear from time to time.

We were not there to witness these events, of course. This Solar Nebula Model is based on computer simulations. A simulation tests a variety of initial parameters until its model evolves into a planetary system much like the Solar System we see today. Running thousands of these experiments increases our confidence that these conditions and processes represent the history of the early Solar System.

In 1995, Swiss astronomers Michel Mayor and Didier Queloz[9] detected a planet orbiting 51 Pegasi,[10] a star like our Sun.[11] This exoplanet, known as 51 Pegasi b [51 Peg b],[12] looks nothing like astronomers expected. It has about half the mass of Jupiter but orbits close to the surface of its host star, taking a mere 4 days to go around it. Since then, the study of planetary systems other than ours has exploded, revealing that many different configurations are possible. Therefore, astronomers continue to explore other models that can produce more exotic arrangements.

A newer theory of the origin of our Solar System, called the Nice[13] Model, posits that the giant planets originally formed much closer to the Sun and then migrated outward. As they did so, they unleashed a flood of comets throughout the Solar System late in the Era of Heavy Bombardment, while also kicking some material outward to form an encircling cloud of comets known as the Oort Cloud and a disk-shaped Kuiper Belt. The latter includes Pluto. ☞ Chapter 14 ☞ In this scenario, more than four giant planets could have formed originally, but a complex game of planetary musical chairs ejected some. With many interactions among the proto-planets and planetesimals, the model could create a variety of planetary configurations based on small changes to the initial parameters. However, so many possible outcomes make the model difficult to test (Figure 1.10).

The Grand Tack Hypothesis suggests that the planet Jupiter originally migrated inwards close to the current orbit of Mars, before interacting with Saturn and reversing its course to its present location.[14] As Jupiter scatters material along its path, it reduces the raw material available for a proto-Mars to accrete and scatters planetesimals into and out of the forming Asteroid Belt. Such movements could produce a transition region between the inner and outer Solar System that more closely resembles the Mars and Asteroid Belt that astronomers observe today. The Grand Tack

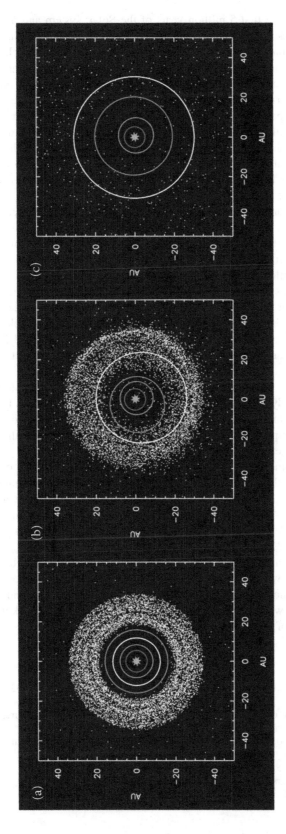

FIGURE 1.10 Computer simulation of the Nice Model. Left: early configuration, before Jupiter (inner circle) and Saturn (second circle from the sun) reach a 2:1 resonance. ☞ Chapter 3 ☞ Middle: scattering of planetesimals into the inner Solar System after the orbital shift of Neptune (outer circle) and Uranus (third circle from the sun). Right: after ejection of planetesimals by planets. In caption, substitute for color codes: green = inner circle, orange = second circle from the Sun, dark blue = outer circle, light blue = third circle from the Sun.

scenario takes place during the first 5 million years of the Solar System and provides the initial conditions for the Nice Model, which is proposed to occur 500 million years later.

Settling on the most probable theory may require new understandings developed through the study of stellar systems beyond our own. This investigation is now in its infancy. With all this exciting work, so much remains that we do not know yet about the history of our Solar System.

Because we live on the Earth and much of its history is written in its rocks, we know more about the later stages in the development of our home planet. The Moon probably formed as a result of a collision between the Earth and a proto-planet during the Era of Heavy Bombardment. ☛ Chapter 4 ☛ During this period, Earth's primary atmosphere of light gases, such as hydrogen and helium, escaped to space. Lasting until about 4.0 billion years ago, this period is known as the Hadean Eon.[15] The molten Earth slowly cooled to form a crust during a time of extreme volcanism. Earth's oldest rocks date to this time. Certainly a world of fire!

From 4.0 to 2.5 billion years ago, the Archean Eon[16] is the time when continental plates formed on the crust and were set in motion. ☛ Chapter 7 ☛ It includes the establishment of Earth's magnetic field that protected its secondary atmosphere, which was mostly carbon dioxide. The first signs of oceans date to this time. Simple life arose early in this period.

The Proterzoic Eon[17] followed from 2.5 billion years to 541 million years ago. Plant life generated significant amounts of oxygen that accumulated in Earth's atmosphere. More complex life forms began appearing at the end of this period. In addition, this eon includes Snowball Earth intervals during which glaciers reached the equator, enveloping Earth in a global ice age. Gale-force winds howled in the cold dry air, far below freezing, and a dark and briny ocean was continually stirred by tides and turbulent eddies beneath the floating ice sheet. Clearly a world of ice!

We live in the Phanerozoic Eon,[18] from 541 million years ago to the present. Although simple life is thought to have arisen nearly 4 billion years ago, most biological evolution has taken place in this eon. Continents drifted apart and eventually coalesced into a supercontinent called Pangea,[19] only to break up and move into their present positions. Geologists project that in 300 million years another supercontinent will come together, Pangea Ultima.

Geologists of the future may find that the Earth bears the definite imprints of our human activities. According to the Anthropocene Working Group in the International Union of Geological Sciences, the professional organization in charge of defining Earth's time scale, a new geological time period has begun. The Anthropocene Epoch[20] can be distinguished within the Phanerozoic Eon by the overwhelming global evidence that humans are altering the Earth's geology, climate, and ecosystems ☛ Chapter 7 ☛.

Our own Earth has been an alien world of fire and, in turn, ice before settling into the comfortable, life-bearing planet we know intimately. The other planets have followed different development paths. We see among them familiar conditions taken to extremes and also completely otherworldly conditions. As we begin to tour our Solar System in detail, we look in the present for clues to our past and hints of our possible future.

NOTES

1 According to the International Astronomical Union in 2009, the exact value is 299,792.458 kilometers per second.
2 Brightest star system in the constellation Centaurus, the half man-half horse creature, according to the 1603 stellar atlas *Uranometria* by Johann Bayer ⟨1572–1625⟩.
3 It is found in the constellation Andromeda. In Greek mythology, Andromeda was an Ethiopian princess rescued from a sea monster.
4 For whom the Hubble Space Telescope, launched in 1990, is named.
5 Slowly varying Cepheid stars have greater luminosities than more rapidly changing ones. The class is named for its prototype, δ Cephei, which is the fourth brightest star in the constellation of Cepheus, the King, according to Bayer.

6 In 2018, the International Astronomical Union recommended this relationship be called the Hubble–Lemaître law, recognizing the observational and theoretical contributions.
7 The pair shared the 1978 Nobel Prize in Physics for their discovery.
8 Named for American cosmologist David Wilkinson ⟨1935–2002⟩, a leading scientist on the COBE and WMAP missions.
9 The pair shared the 2019 Nobel Prize in Physics for their discovery.
10 The 51st star in the constellation Pegasus, the winged horse, based on the order they were listed in the star catalog *Catalogus Britannicus* prepared by English Astronomer Royal John Flamsteed ⟨1646–1719⟩.
11 For this discovery, the duo shared the Nobel Prize in Physics in 2019.
12 The first planet ("b") discovered orbiting the star, 51 Pegasi.
13 Pronounced "niece" after the French city where the theory originated.
14 Its name is an analogy to a sailboat tacking into the wind.
15 From the Greek god of the dead and the underworld.
16 From the Greek *archaîos*, meaning ancient.
17 From Greek *próteros*, meaning prior.
18 From Greek *phanerós*, meaning visible.
19 From Gaia, the Greek goddess personifying Earth.
20 An epoch is a subdivision of geologic time, within an eon.

2 Characterizing Planets

To our earliest ancestors, the planets were points of light in the night sky, differing little from the multitude of stars above. Most stars remain in the same place with respect to each other. That is, they seem to be mounted upon a Celestial Sphere forever whirling about us. Early sky watchers named the apparently permanent patterns of stars on the imaginary Celestial Sphere; these developed into the familiar constellations we identify today. However, still in pre-history, people noticed that seven 'stars' *moved* with respect to all the other so-called fixed stars, in that they wandered through some of the constellations. They were named 'planets,' the Greek word for wanderer. This definition of planet as wandering star included the Sun, the Moon, Mercury, Venus, Mars, Jupiter, and Saturn (Figure 2.1).

Most of ancient astronomy was taken up with attempts to understand the motions of the planets as well as the apparent motions of the Sun and the Moon. The classical Greek scholar Eudoxus ⟨*circa* 400 BCE⟩ came up with an ingenious plan in which nested, concentric spheres surrounded the Earth. Upon each sphere was mounted the Sun, Moon, or another planet. The outermost sphere was the home of the stars. The key ingredient to Eudoxus's scheme was that all but the last sphere were transparent, made of some substance like crystal. You could see through them. Each planet's sphere rotated at its own particular speed upon its own individual axis, which generated complicated relative motions played out against the background stars. Seemingly elegant, his system could make only crude predictions of where the Sun, Moon, and other planets might be on a given day (Figure 2.2).

The Hellenistic astronomer Claudius 'Ptolemy' Ptolemaeus ⟨*circa* 150⟩ thought he could do better. He was especially keen on solving the mystery of retrograde motion, times when certain planets appeared to pause, back up (move westward), pause again, and then continue their typical (eastward) motion with respect to the stars. This pattern is particularly noticeable in the orbits of Mars, Jupiter, and Saturn. Ptolemy used a deferent, a circle around the Earth upon which was mounted, not a planet, but another smaller wheel called an epicycle. The planet was attached to the epicycle. By giving the deferents and epicycles their own rates of uniform circular motion, Ptolemy could create retrograde motion and better predict the positions of all the bodies he thought revolved about the Earth (Figure 2.3).

Thanks to Polish astronomer Nicolaus Copernicus ⟨1473–1543⟩, we no longer believe that everything revolves about the Earth.[1] He theorized that the Sun should have been the center of Ptolemy's deferents, not the Earth. Indeed, the Earth was just one of six planets orbiting the Sun. Only the Moon revolved around the Earth. In the Copernican model, retrograde motions occur naturally when the Earth overtakes a slower moving planet, causing it to appear to move backward for a time, as it recedes from view (Figures 2.4 and 2.5).

We have come a long way from Eudoxus and Ptolemy. The Earth is not the center of the Solar System. The planets and stars need not be mounted on anything, but rather move through space on their own.

The modern study of the planets began with the work of German astronomer Johannes Kepler ⟨1571–1630⟩ in the 17th century (Figure 2.6), who theorized that all the planets revolve about the Sun in elliptical orbits with the Sun at one focus; the other is empty. This is usually called Kepler's First Law. The orbital eccentricity describes how 'flattened' the ellipse is, varying from 0 for a perfect circle and approaching 1 for a highly elongated figure. The longest distance between two points on an ellipse is its major axis; all points on a circle are at an equal distance, its radius. Among solar-system planets, Venus has the most circular, or least eccentric, orbit. In contrast, Mercury's orbit is the most squashed with an eccentricity of 0.2.

FIGURE 2.1   Planets Saturn and Mars against the background of stellar constellations.

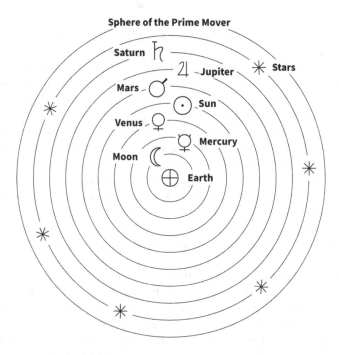

FIGURE 2.2   Eudoxus's concentric spheres.

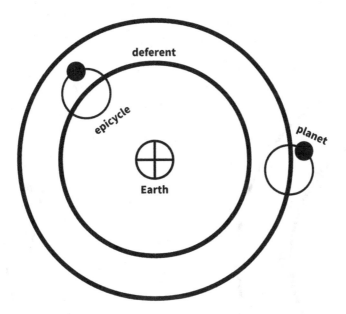

FIGURE 2.3 Ptolemic Model.

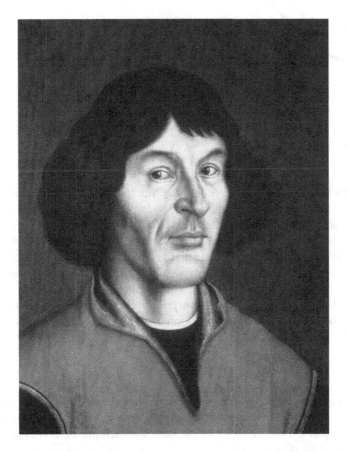

FIGURE 2.4 Oil portrait of Nicolaus Copernicus.

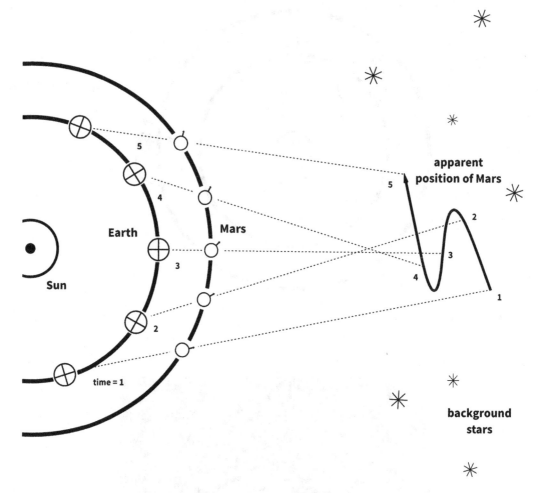

FIGURE 2.5   Retrograde motion explained in the Copernican system.

Moreover, planets travel fastest when nearest the Sun (perihelion)[2] and slowest at their farthest point from the Sun (aphelion), half an orbit later. The implication is that most of a planet's orbital time is spent around aphelion. This is called Kepler's Second Law (Figure 2.7).

Kepler gave us the Solar System we know today: Planets, including the Earth, revolving about the Sun more-or-less in a plane called the ecliptic. The Solar System is 'flat.' Planets revolve counterclockwise as viewed from the direction we on the Earth call North. However, which planet orbits where? Specifically, how far[3] is each from the Sun?

Kepler's Third Law is arguably his most powerful because it relates the planets to each other. Kepler found that the period of an orbit (time) of revolution [T] of a planet is proportional to half of its major axis or its semi-major axis [a], but in a precise way: Kepler came up with the relation

$$T^2 = N \cdot a^3$$

where N is a constant and depends upon the units of time and distance used (e.g., hours, years, *etc.*; kilometers, miles, *etc.*). Planets farther from the Sun move more slowly than those closer. Whatever N is—Kepler did not know—it is the same for each planet. It is kind of like the scale factor on a model car kit (Figure 2.8).

FIGURE 2.6 Oil portrait of Johannes Kepler.

What is called Kepler's Third Law is predictive. If you know the semi-major axis, you also know the orbital time. More useful, if you measure T by watching the planet revolve about the Sun and noting the time that is required to return to the same place with respect to the Sun, you can calculate a. Knowing the orbital periods of the planets, Kepler could determine that, if planet Mercury was so far from the Sun, slower planet Saturn must be at another, greater distance. And so, the planets were arranged in the familiar elementary-school litany: Mercury, Venus, Earth, Mars, Jupiter, and Saturn. Although measuring the true, physical distances between the planets and the Sun would take centuries, knowing the relative distances began to shape our understanding of the Solar System thanks to Copernicus and Kepler (Figure 2.9).

English physicist Isaac Newton ⟨1642–1727⟩ demonstrated that all of Kepler's Laws were a consequence of his theory of gravitation (Figure 2.10): If gravity exists, planets simply must move as Kepler's Laws say they do. ☛ Chapter 3 ☛ Newton improved the Third Law to include the planets' relative masses. This important revision now makes measuring mass possible, specifically, the total mass of any bodies orbiting one another—like a cosmic balance.

With the increasing acceptance of a Sun-centered, or heliocentric, Planetary System, naked-eye observers tracked the motions of five planets through the night sky. The Earth brought the total of planets to six. Then in 1781, German-English musician-turned-astronomer William Herschel ⟨1738–1822⟩ was scanning the skies with a telescope that he had built himself. He came upon a "fuzzy" star (Figure 2.11). At first, he thought that he had spied a new comet—pleasing but one of many and not that unusual an occurrence. Instead, it turned out to be an epoch-making event.

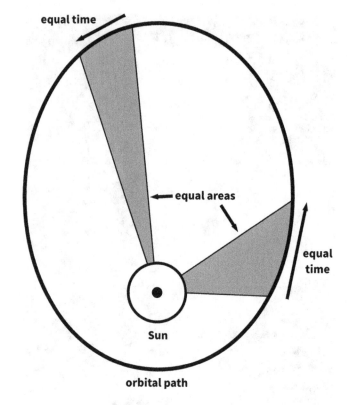

FIGURE 2.7   Kepler's First and Second Laws (not to scale).

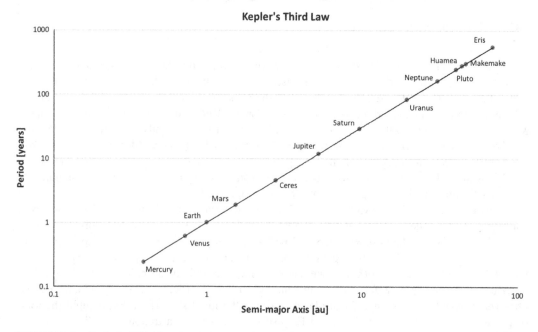

FIGURE 2.8   Kepler's Third Law.

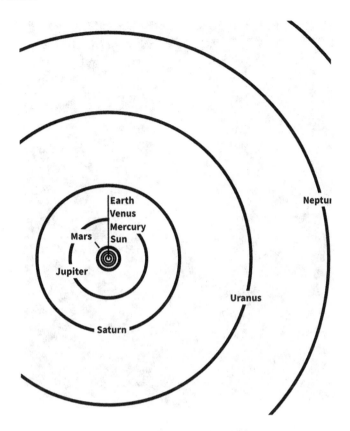

FIGURE 2.9   Planet orbits to scale.

Herschel had come upon a seventh planet, now known as Uranus,[4] orbiting the Sun beyond Saturn and confirming Kepler's Laws. This was the first planet to be discovered using a telescope. ☞ Chapter 11.

Giuseppe Piazzi ⟨1746–1826⟩, an Italian astronomer and priest, detected Ceres,[5] another new planet in the gap between Mars and Jupiter. At the time, the telescopic discovery of new planets was not surprising. While Ceres was too small to show a disk, its characteristic 'wandering' against the background of stars revealed its presence. Its orbit was beyond that of Mars but within that of Jupiter, it orbited the Sun approximately in the plane of the ecliptic, and it reflected relatively little sunlight—all things you might expect from a trans-martian planet.

The apparent expansion of the Solar System from seven to eight planets would have been the end of the story if another new planet had not been discovered moving in a similar orbit! The 1802 discovery was named Pallas.[6] Soon astronomers were filling the once empty gap between Mars and Jupiter with similar bodies: Juno[7] in 1804, Vesta[8] in 1807, Astraea[9] in 1845, and so forth. The region rapidly became crowded with these.

Herschel compared and contrasted his observations of Ceres and Pallas with the known characteristics of established planets and comets. He found that the two new worlds did not fit well with either category and, so, concluded that they represented a new type of solar-system object, which he chose to call asteroid, or 'star-like.' In using asteroid for the proposed category, he emphasized their star-like appearance in telescopes of his day; he could not resolve either into a clear disk as he could the previously known planets. Indeed, Ceres is only 950 kilometers (590 miles) across—not very big by planetary standards. The region two to three times the Earth's distance from the Sun, in which most of these bodies orbit, eventually became the Asteroid Belt.

FIGURE 2.10   Oil portrait of Sir Isaac Newton.

Over the next 200 years, astronomers added Neptune[10] and Pluto[11] to the ranks of planets. Then, in 2003, a team led by American astronomer Michael Brown discovered Eris,[12] a solar-system body roughly the size of Pluto but slightly more massive. The International Astronomical Union [IAU] responded in 2006 with a formal definition of 'planet' and 'dwarf planet.'

In modern terms, a planet orbits the Sun, has sufficient mass to organize itself gravitationally into a nearly round shape, and has cleared the region surrounding its orbit. A dwarf planet is similar but has not cleared its neighborhood. According to the IAU, all other bodies orbiting the Sun directly are "Small Solar System Bodies." Therefore, we have eight planets: Mercury, Venus, Earth, Mars, Jupiter, Saturn, Uranus, and Neptune. Most of these planets also have bodies orbiting them; these objects are formally called satellites but more commonly known as moons.

Ceres and Pluto are reclassified as dwarf planets along with Eris, Haumea[13] (2003), and Makemake[14] (2005). Although, as round bodies orbiting the Sun, they have planet-like characteristics, none of these worlds is the dominant system in its region. Moonless Ceres is the largest member of the Asteroid Belt while Pluto and its five moons are part of the Kuiper Belt of icy, rocky objects orbiting beyond Neptune. ☞ Chapter 14 ☞

Even as the definition of planet was evolving in response to new discoveries, astronomers were also using different techniques to establish the scale factor for the Solar System, Kepler's N. In the 18th and 19th centuries, scientific expeditions traveled to observe Venus pass across the face of the Sun, or transit, in order to calculate the distance between our planet and the Sun. ☞ Chapter 6 ☞ Astronomers now know this distance to be about 150,000,000 kilometers[15] (93,000,000 miles), or roughly as far as all the vehicles on Earth drive in a year (Figure 2.12).

And those distances are staggering. Measuring the distance between planets in kilometers would be akin to measuring the distance to Tokyo in inches—too many zeros! So, astronomers use their

FIGURE 2.11 Lithograph Portrait of Sir William Herschel.

own distance unit, the Astronomical Unit [au], and set it equal to one of the best physical distances modern astronomy has derived: the average distance between the Sun and Earth. Mercury and Venus are less than an au from the Sun, the Earth is exactly 1 au, and the rest of the planets are several au away. Keep in mind that our Sun's gravity binds objects, mainly comets, to it at distances of tens of thousands of au, about one-third the distance to the next closest star. That is why we refer to the inner region in which the planets inhabit as the Planetary System as opposed to the Sun's whole sphere of influence, the Solar System.

The au is convenient, but is it not arbitrary? Yes, it is! But at one time people had to decide that the distance from someone's nose to outstretched fingers was a 'yard' and whose foot '1 foot' was. The meter was originally 1/10,000,000[th] of the distance from the Equator to the North Pole. All these distance units are, in a sense, arbitrary.

When we place the planets at their proper distance from the Sun, they neatly divide into two groups: an inner Planetary System, those planets, Mercury through Mars, that huddle close to the Sun as if for warmth and an outer Planetary System, those frigid planets, Jupiter through Neptune, much farther from the Sun.

When confronted by a new phenomenon, scientists often start by classifying examples. By sorting the planets based on various properties, we see at once how they differ and how they are the same.

Once we know how far away a planet is, we can use its apparent size through the telescope and calculate its physical size (diameter). The result? The inner Planetary System is populated by comparatively small planets, while the planets of the outer Planetary System are giants.

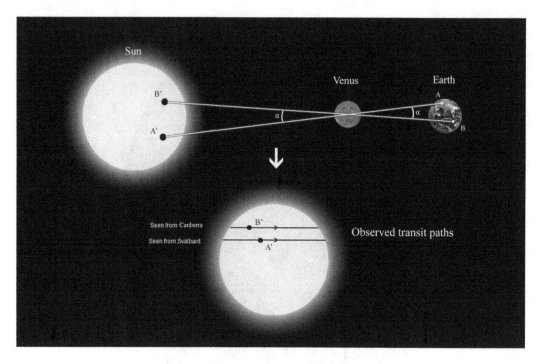

FIGURE 2.12 Path of planet Venus, as it transits the Sun, seen from two different locations on the Earth. The angle between the two can be used to determine the length of the astronomical unit.

What common measurements are taken when you visit your doctor's office? The answer is height and weight. The 'height' of a planet is its diameter. The 'weight' (more properly, mass) of a planet can be obtained by watching another object orbit about the planet. Gravity is a function of mass, and an orbiting object (a satellite, for instance) at a given distance from the planet must have a certain mass to neither fall into the planet nor escape it altogether. To no one's surprise, the bigger planets are also more massive than the smaller planets when we apply Newton's cosmic balance (modified Kepler's Third Law).

Looking at the scaled diameters of the planets, you see the outer planets dwarf those of the inner planets. However, we do not appreciate the true difference in size until we think of the planets in three dimensions (Figure 2.13).

Imagine that the first planet, Mercury, is the size of a pea. Venus and the Earth would be the size of grapes, and Mars would be something like a blueberry. Why are we scaling the planets so small?

FIGURE 2.13 Relative sizes of the planets, Pluto, and Sun (distances not to scale).

We need to, because of what is going to happen next: Into this 'solar-system salad' we now add Jupiter—a cantaloupe. Saturn is the size of a grapefruit, and Uranus and Neptune are a lemon and a lime. This analogy is conservative (and seasonal!); the difference in sizes of the planets are greater than these props, but the point is made. Jupiter, Saturn, Uranus, and Neptune are gigantic compared to Mercury, Venus, the Earth, and Mars. The Sun to this grocery-store scale would be the biggest, prize-winning pumpkin you have ever seen.

Notice how we are building a ladder of properties of planets? From simple geometry, we can now calculate a spherical planet's volume, based on its diameter. Mass divided by volume gets us the density of the planet, a most important parameter.

Imagine we put a box of lead fishing weights in the right pan of a balance scale and some cotton in the left. Which way will the balance tilt? The answer is *left*, toward the cotton. Something seems wrong here, and it is that we have left out an important piece of information: It was a whole bale of cotton.

This does not seem fair. We know that, had we put the cotton in the same-sized box as the lead, the balance would tip right, toward the lead. This is because it is a characteristic of cotton that, all else being equal, it will have less mass than lead.

Astronomers find the average planetary density in the inner Solar System to be about 4 g/cm^3; in comparison, water has a density of 1 g/cm^3. We then look for materials that have densities near this value. Substances denser than water are not hard to find. Rock and metal have densities on this order. We then make the following leap: These planets may be made of rock and metal with little or no atmosphere. We live on one of those planets and know that it is indeed made of rock and metal. While the Earth's atmosphere is just right for us, it is thin compared to those of the outer planets. Worlds like ours are called terrestrial, or Earth-like, from *terra*, which is Latin for Earth.

If we calculate the average planetary density for the planets in the outer Solar System, we get an interesting result: a lower average density of a little more than 1 g/cm^3. Rock and metal cannot be the primary components of these worlds. They must be made of something else. All have thick atmospheres made up of the lightest gases and extensive liquid/metallic hydrogen interiors with a relatively small rocky, metallic core. These are giant planets and, by far, the greatest of all of them is Jupiter. Indeed, Jupiter is representative of the four outer planets, which are also called Jovian Planets, after an archaic name for Jupiter, Jove. We can further divide these behemoths into Gas Giants (Jupiter and Saturn) and Ice Giants (Uranus and Neptune). Density is such an important way of classifying planets because it begins to let us know their intrinsic nature.

The brightness of a planet can also indicate its surface features. Given a planet's size and distance from the Sun, we can calculate how much sunlight strikes it. Astronomers also can measure how much of that light is reflected towards us. The fraction of light reflected to sunlight received is known as albedo. A perfect reflector would have an albedo of 1 while something that absorbs all light would have an albedo of 0. Enceladus, an icy moon of Saturn, reflects sunlight effectively with an albedo of 0.85. However, different conditions can produce similar albedos. Venus also has a high albedo (0.75) due to its bright cloud layer. The Moon, however, is less reflective with an average albedo of 0.11 due to dark basaltic plains on its surface.

We can do better than just calculate the average properties of a terrestrial planet. Geophysicists can probe its interior. Chinese scholar Zhang Heng ⟨78–139⟩ developed an 'earthquake weathercock' to monitor even slight tremors. Starting in 1875 with an instrument designed by Italian physicist and priest Filippo Cecchi ⟨1822–1887⟩ to record earthquakes, a world-wide network of seismometers now detects tremors moving through the Earth. Planetary scientists also have placed seismometers on the Moon and Mars. In addition, they are beginning to detect and interpret vibrations in Jupiter.

In a seismic event (e.g., an earthquake or moonquake) mechanical energy is released in the form of waves. Two kinds of waves are generated: primary, pressure [P] waves and secondary, shear [S] waves. The P waves travel more quickly than S waves and are the first to arrive at a seismometer. To model a P wave, find a toy Slinky®[16] and compress a few rings on one end. When you release this high-density bunch of rings, the compression will travel to the other end of the Slinky as a single wave. To model a S wave, have a friend help you hold a Slinky taut along the floor. Then, yank it sideways near one end. The resulting bump will travel to the other end of the coil as a single wave (Figures 2.14 and 2.15).

The speed of P and S waves depends on the material through which they travel. P waves pass more quickly through dense rock than liquids. S waves do not travel through liquids or gasses at all

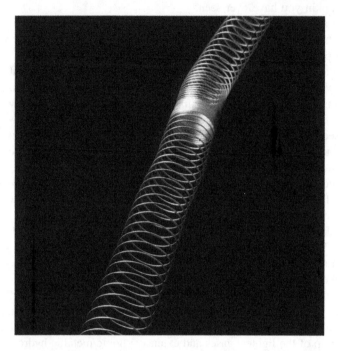

FIGURE 2.14   P wave in a Slinky.

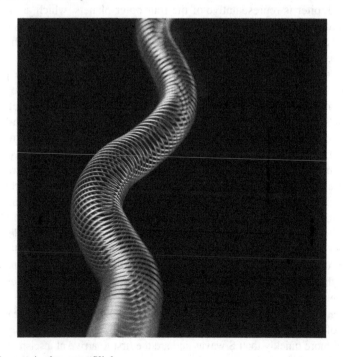

FIGURE 2.15   S wave in the same Slinky.

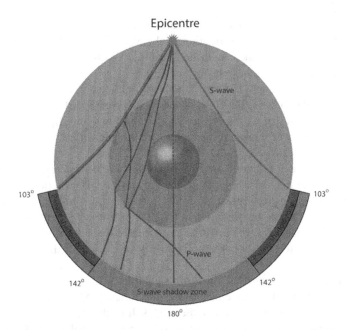

FIGURE 2.16 Seismic waves travel through a planet. S waves do not penetrate a fluid layer.

because such substances lack rigidity; that is, you can grab the end of a skipping of rope and generate an S wave, but you cannot grab the water streaming from a garden hose and do so. By detecting changes in the speed and direction of these waves as they travel through a body, the boundary depths of different layers can be determined. Seismic studies allow us not only to estimate the composition of a planet layer-by-layer, but also the state of each layer (Figure 2.16).

In addition to orbiting the Sun and shaking internally, planets turn, or rotate, about their own axis. Rotation contributes to the weather systems that develop in planetary atmospheres.

Finding the rotation period of a planet with surface features is straightforward: Pick a distinctive feature and wait until the planet rotation returns it to the same location. The interval is one rotation period. The time between sightings can be adjusted to account for simultaneous motion of the planet around the Sun and the Earth relative to both.

In the 1930s, several countries began developing systems to detect aircraft and ships by broadcasting radio waves[17] and interpreting any signal bounced back; such systems became known as 'radar,' for RAdio Detection And Ranging. Continued improvements in the technology after World War II led to its use in detecting a wider range of phenomena, such as the rotation rates for planets and their surface features.

The rotation period of a planet reckoned with respect to the Sun yields the familiar 'day,' which, on Earth, we measure from midnight to midnight or sunset to sunset or so forth. This is the planet's synodic rotation period. Depending on all the relative motions, a planet may turn more than 360° around its axis in order to complete a full synodic rotation; it also may turn less than 360°.

If, instead, the planet's rotation is measured with respect to more distant stars, it will turn a full circle and just 360°. This is its sidereal[18] rotation period. A planet's sidereal rotation period does not always equal its synodic period.

The Terrestrial Planets have rotation periods of spin that range from 1 day for the Earth to 243 days for Venus. For all their size and mass, the Jovian Planets rotate quickly, with the slowest, Uranus, taking just over 17 hours. Because their outer layers are gaseous, the rotation rates of the Jovian Planets vary from equator to pole.

Excepting Venus and Uranus, the planets turn on their axes in the same direction as they orbit the Sun: counterclockwise as viewed by an astronaut far above the North Pole. Similarly, most moons orbit their hosting planets in the same direction. The planets and their natural satellites retain this common motion from the formation of the Solar System while deviations hint at collisions and captures that changed the established patterns. Venus and Uranus rotate backwards, or rotation - spin, compared to the other planets.

The amount by which the angle between a planet's rotational axis and its orbital axis deviate from perpendicular is its obliquity. Mercury, which rotates nearly vertically, has an obliquity near zero. The seasonal variation that we experience on Earth is due to the slightly more than 23° tilt of the Earth's rotation axis. The retrograde rotators, Venus and Uranus, have obliquities greater than 90°.

Planetary composition and rotation conspire to produce another interesting distinction between the planets: Some act like magnets. Spinning electrical current will generate a magnetic field. For instance, electricity from a battery passing through coils of wire spiraling about an iron nail will behave like a magnet. A conducting fluid within a rotating planet will do the same thing, a process known as a 'magnetic dynamo.' The Terrestrial Planets have weak or no magnetic fields compared to the strong fields generated by the metallic hydrogen layers within the rapidly rotating Jovian Planets (Figure 2.17).

Representing an abstraction such as a magnetic field that we cannot see is difficult. To do so, illustrators draw magnetic field lines emanating from a planet. Magnetic and charged particles, such as the solar wind, align themselves with these lines. It is like the pattern created by iron filings sprinkled over a bar magnet. These lines converge at the magnetic poles, which are not the same as the rotational poles. When charged particles moving along these lines interact with atoms in the atmosphere, they produce light known as the aurora.[19] Once described metaphorically as a 'northern dawn' (*aurora borealis*) on Earth, astronomers have detected aurorae on the giant outer planets as well.

Summing up so far, here are the properties that make planets differ from one another:

1. distance from the Sun (temperature)
2. diameter (size)
3. mass
4. density
5. albedo (reflectivity)

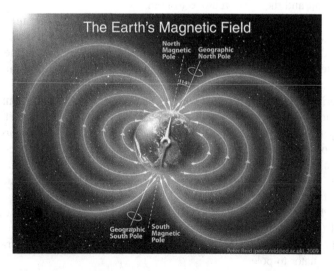

FIGURE 2.17 Planetary magnetic field lines. Notice that the magnetic poles do not necessarily correspond to the geographic poles.

6. composition
7. rotation and obliquity (tilt)
8. magnetic field

In the next chapter, we will discuss the orbital properties of planets and other solar-system bodies.

In addition, we can use these properties to evaluate the moons, or natural satellites, orbiting individual planets. The Terrestrial Planets host few or no moons while the Jovian Planets are home to complex systems of moons and rings. Some of the bodies orbiting Jovian Planets are as large as Terrestrial Planets.

The Solar System also contains a variety of smaller bodies orbiting the Sun, which range from interplanetary dust to meteoroids to asteroids and comets to dwarf planets. Once again, we can use the same basic properties to study these minor bodies.

We have accumulated most of the tools we need to study the Solar System from the comfort of our armchairs. We have a formal definition that identifies a planet from other bodies and have identified the fundamental characteristics that describe a world, whether planet or other. While we have more techniques and greater rigor than our early ancestors did, we have the same capacity to marvel at the wonders revealed.

NOTES

1 The Greek astronomer Aristarchus of Samos ⟨*circa* 230 BCE⟩ also proposed putting the Sun in the center but had less success in convincing his colleagues that the Earth moved.
2 From Helios, the Greek god of the Sun.
3 Because planetary orbits are ellipses, a planet's distance from the Sun varies during each orbit. Therefore, distances quoted are average distances.
4 Greek god of the sky; father of Roman Saturn.
5 Roman goddess of corn and harvests.
6 Another name for Athena, Greek goddess of wisdom and war.
7 Roman queen of the gods, protectress of marriage and women.
8 Roman goddess of the hearth.
9 Greek goddess of justice.
10 Roman god of the sea.
11 Roman god of the underworld, Hades in Greek mythology.
12 Greek goddess of discord, best known for igniting the Trojan War.
13 Hawai'ian goddess of fertility.
14 Rapa Nuian god of fertility.
15 In 2012, the IAU defined it exactly as 149,597,870.700 kilometers.
16 Any spiral spring will do, but we still play with Slinkys.
17 Radio waves are a form of electromagnetic radiation, or light, with wavelengths longer than 1 millimeter (0.039 inches). Radar systems frequently operate in the 0.3–30 centimeter (0.1–10 inch) range.
18 'Sider-' is a prefix meaning 'star.'
19 Roman goddess of dawn.

3 Maintaining Orbits

Modeling the motion of the planets challenged astronomers for more than two millennia. We have considered the crystalline spheres of Eudoxus, the deferents and epicycles of Ptolemy, and the perfect circles of Copernicus before settling on ellipses as described by Kepler. However, Kepler did not understand the force that moved the planets and attempted to explain their separations in terms of regular solids, such as a sphere, a cube, and a dodecahedron.[1] Once Newton provided the concept of gravity that holds the system together, all the pieces that make up our modern understanding of planetary motion fell into place.

The distance of a planet and its motion about the Sun—in short, its orbit—are key to almost every aspect of a planet. Its orbit defines what kind of planet it is, how long it will last, how it will appear to us, and how it will behave throughout its lifetime; similarly, orbits describe the behavior of lessor bodies down to grains of interplanetary dust. ☞ Chapter 9 ☞ Planets execute the sort of perfect figure in space of which Olympic skaters can only dream. A planet's orbit is truly a thing of beauty.

Indeed, the physics of orbits gives movement to the whole Solar System: planets, asteroids, and comets around the Sun or moons and ring particles around their home planet or the Solar System around the center of the Milky Way Galaxy. Therefore, we will take a few pages to explain how orbits work. Although orbital mechanics is mathematically elegant, we will restrict our discussion to a qualitative description of orbits, using planets as the principal example. While refraining from using a single equation, we will answer the following question: How do those things stay up there, anyway?

Newton explained his discoveries in terms of "standing on the shoulders of giants." To understand planets and planetary orbits, we, too, must begin by investigating two foundational ideas. First, we need to look at motion itself. How does a thing—anything—move? Second, we need to look at a force called gravity.

In our busy modern age, motion seems natural. Is it? Aristotle ⟨384–322 BCE⟩, the Greek thinker who established the tradition of Western science, believed not. He was not very impressed with motion. He felt that rest was the natural state of all things (Figure 3.1).

Italian Galileo Galilei ⟨1564–1642⟩, the Renaissance polymath, thought differently. He questioned whether either motion or rest was a 'preferred' state of objects. He did not think that there was anything more 'natural' about rest than about motion, or vice versa. By eliminating preferred states, Galileo inaugurated the modern study of motion called dynamics (Figure 3.2).

Newton picked up where Galileo left off. He stated that an object at rest remains at rest and that an object in motion remains in motion—traveling in a straight line, at a constant speed. This is known as Newton's First Law of motion.

The only way to put an object at rest into motion, stop a moving object, change the speed of an object, or change the direction of a moving object is to exert a force, a push or a pull, on it. According to Newton, the change in motion, produced by a force, occurs in the direction of the force and is proportional to its strength. This is Newton's Second Law.

These laws make sense. When you pick up and roll a bowling ball, it travels forward in the direction you released it, not over your shoulder or somewhere else. Furthermore, the harder you throw it, the faster the bowling ball goes.

What is a 'change in motion'? Scientists combine the speed and direction of a moving object into a single concept, velocity. A change in velocity is an acceleration: speed up, speed down, or turn. So, a force produces an acceleration, which is a change in motion (Figure 3.3).

We usually think of an acceleration as an increase in speed. It is. But, it can be, just as easily, a decrease in speed although we tend to call such a change a deceleration. We are less likely to think

FIGURE 3.1 Bust of Aristotle.

of a change in direction as an acceleration. However, a change in direction is just as much a change in velocity as a change in speed is. A force can produce a change in speed, a change in direction, or both—depending on how the force is applied.

Newton was on a roll and kept going. Acceleration is proportional to the force applied, he declared. It also is proportional to a property of every object, its mass. Mass is simply a measure of the amount of matter within an object. Given a certain force, if you try to change the motion of a large mass, you will get less acceleration than you would with a smaller mass. Even pushing very hard, it takes much effort to get a stalled car rolling. Yet, what happens when you use the same amount of force on a skateboard? Whoosh! A very rapid change in velocity results.

Completing the trifecta is Newton's Third Law. For every action, there is an equal and opposite reaction. For example, if you row a boat, your paddles move backwards, propelling the boat forward. This also applies to the launch of a spacecraft. Hot exhaust gases are expelled from the end of the rocket causing it to rise off the pad and climb into the sky.

With these three laws, any general motion can be described: a ball thrown to a friend, a flying aircraft, or a race car speeding around a track. These laws are powerful and extremely accurate. NASA uses them to dispatch spacecraft to distant destinations in the Solar System.

Newton showed that motion is not capricious, that it follows relatively straightforward and describable rules. Newton's Laws never fail us. You can bank on them because they cannot be broken. They are also vital to understanding the orbits of planets.

FIGURE 3.2   Oil portrait of Galileo Galilei.

FIGURE 3.3   A speedometer measures speed. A compass indicates direction. Combined, they provide the velocity of a car.

Gravity is both simple and complex. In a sense, we all know what gravity is. Nevertheless, you could take an entire graduate physics course on gravity without grappling with all its nuances.

For our purposes, gravity is simply a force, but a special kind of force. First, it is always a pull and never a push. Second, unlike most other forces, it can act at a distance. In our earlier example of acceleration, you had to place your hands on the rear bumper of the car to get it moving, but gravity can accelerate an object without any direct physical contact.

Gravity is produced by any object that has mass. As everything that we know of (physically) has mass; this includes the whole Universe of objects! Between each pair of these objects there is the attractive force of gravity along the line joining the two—the more mass in the pair of objects, the stronger the force. Think of that. Right now, you (an object with mass) and the distant planet Neptune (another object with mass) are attracted to each other. Do you feel it? No, of course not. That is because gravity grows weaker the greater the distance between objects (more precisely, the square of the distance). It becomes infinitesimally weak at infinite distance, but never zero; we all are bound together in this Universe.

To experience the force of gravity, you need only leap upward. The velocity that you give yourself with your feet is quickly decelerated away. There is a moment of 'hang time,' and then you are accelerated down toward the massive object underneath you, the Earth. You do most of the moving because the Earth is much more massive than you. Technically, the gravity produced by the mass of the Earth acts as if it is emanating from the center of the Earth, some 6,400 kilometers (4,000 miles) below you. Before you get there, though, another force, that exerted by the hard floor, stops you (Figure 3.4).

Now back to our original question: With all those bodies revolving around in the heavens—planets, moons, comets, and asteroids—why does everything not fall down?

For centuries, people wondered what might propel the planets in their paths. Some speculated that it might be a magnetic force, magnetism being a popular phenomenon to dabble with in the physics of the 17th century. Then, Newton described gravity and explained that the same force that works in day-to-day life on the Earth also affects the planets. An apple falling from a tree, a man pushing a shopping cart, the Moon orbiting the Earth, or a comet approaching the Sun all follow the same laws of motion; their movement is consistent with a force that is proportional to the masses involved and inversely proportional to their separations. Newton's immortality rests in his realization that the force that runs the Solar System lies—literally—beneath our feet (Figure 3.5).

FIGURE 3.4 Hang time.

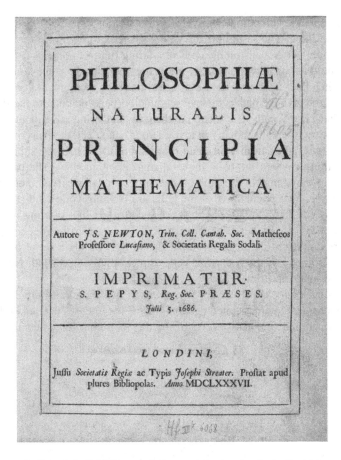

FIGURE 3.5   Newton's *Principia* (1687), one of the most famous books in the history of science.

Rather than keeping objects in the sky, this force requires that they do the opposite—fall toward each other! An attraction exists between the Earth and its Moon, between the Sun and each of its planets, and between all other objects as well: Gravity is this attractive force.

"Why do the planets not fall?" we asked. They are falling—all the time! They must be. Recall that in the absence of a force, an object is either at rest or moving at a constant speed in a straight line. That goes for any object, anywhere in the Universe. With no force acting upon it, a planet would either be stationary or traveling at a constant rate toward or away from us. If the latter were the case, the planet would either crash into something or disappear forever in the distance. No real planet behaves in this manner. Planets move roughly in nearly circular ellipses with their speed and direction of travel perpetually changing.

If gravity is the force acting on the planet—and it must be—should every planet collide with the Sun? This would be the case, if the planet started at rest. In other words, if you placed a new planet at rest with respect to and some distance away from the Sun, imparted no other force on it, and let it go, the planet would fall into the Sun. No question.

What if the planet is not only attracted by the gravity between it and the Sun, but also has some initial velocity? If the direction of this velocity is toward or away from the Sun, it will not make any difference ultimately, but what if it is not?

We say that the direction toward or away from a center is radial. For instance, spokes of a bicycle wheel are aligned radially. The direction perpendicular to this is transverse. A planet that starts with a transverse velocity will be acted upon by the radial gravitational force of the Sun, perpendicular to the planet's direction of travel. Its direction of travel will then change; the planet is accelerated by the force. The resulting path of the planet will be a curve.

The classic example is two balls on the ends of a string (part of *boleadoras*[2] in certain parts of the world). If you hold the balls motionless by the string, they dangle to the floor. If you give them a radial velocity (pitch them), they move in a straight line (for a moment, at least). However, if you hold onto the string and give the balls a transverse velocity (perpendicular to the string) by twirling it, the balls travel in a curved path. The tension in the now-taut string takes on the role of gravity, in this analogy, by drawing the balls radially in toward your hand. Yet, the balls do not move inward. They stay at a constant radius from your hand (the length of the string). Although their speed is constant, the direction in which the balls move changes continuously. The balls continually accelerate, but never hit anything. In fact, they travel in a particular curved path that repeats itself over and over: a circle. You could say that the balls are orbiting your hand (Figure 3.6).

To be sure, Newton's Laws still apply. Just cut the string while twirling it and look at what happens. The balls take off. Rather than move radially outward from you, they continue in the direction they were pointed at the instant you cut the string. Aiming a set of *boleadoras* requires timing your release to get just the right transverse direction. We need both Newton's Laws and Newton's gravity to make an orbit.

An object traveling a curved path really is moving in two directions simultaneously: radially and transversely. Consider a giantess pitching a ball. Once she releases the ball, it flies at a constant speed parallel to the ground (ignoring effects of the air). No other force acts upon the ball in the

FIGURE 3.6 Native American hunting with *boleadoras*. Copper plate engraving.

transverse (horizontal) direction. In the radial (vertical) direction, however, gravity drags the ball toward the ground. It accelerates radially until it hits the turf; in other words, it falls downward. The two motions exist independently of each other. The ball could have continued traveling at the same transverse speed indefinitely, had its radial motion not caused a collision with the ground (Figure 3.7).

What if the giantess wants to throw the ball farther? She could instead pitch it radially. This time, as the ball leaves her hand, it travels upward. Gravity must now decelerate this upward velocity to zero before accelerating the ball downward to *terra firma*. The ball takes much longer to strike the ground. Unfortunately, without any transverse velocity, the ball makes no progress across the field and falls at her feet (Figure 3.8).

The giantess's best bet is to throw the ball at an angle to the ground so that it has both a transverse and a radial velocity. The radial component 'buys time' for the transverse velocity to gain yardage. Still, no matter what force the giantess imparts on the ball, it will always 'bite the dust' eventually (Figure 3.9).

Suppose our giantess is so tall that she can see the curve of the Earth. If she hurls the ball hard enough, as the ball falls toward the ground, the ground curves away beneath it. The ball maintains its transverse velocity. It continues to be pulled downward (meaning toward the center of the Earth), but now that direction has changed. As the ball travels parallel to the ground, it continues to fall downward as the ground continues to curve away from under it. If the giantess has given the ball a sufficient initial transverse velocity, she had better look out! The ball will 'fall' all the way around the spherical Earth and hit her in the back of her head. If she ducks, it will continue to travel around the Earth, over and over; it is now in orbit about the Earth (Figure 3.10).

When we watch the flames and clouds of smoke pouring forth from the engines of a rocket on its launching pad, all that thrust serves only one purpose: It is to impart on the rocket a transverse velocity great enough so that the spaceship atop it will achieve orbit (Figure 3.11).

A particular orbit requires a precise combination of gravity and transverse velocity. Is it a coincidence that so many planets, moons, and other bodies achieved this necessary combination? Not really. First, many combinations can produce a stable orbit; hence, the thousands of artificial satellites orbiting the Earth. Second, we are seeing just the result of billions of years of solar-system evolution. Only those objects with the right velocities survived so that we can see them today. How many planets did not, and ended up falling into the Sun or were expelled? (Figure 3.12).

FIGURE 3.7 Giantess throws a ball transversely.

FIGURE 3.8 Giantess throws the ball radially.

FIGURE 3.9 Giantess throws the ball at an angle with respect to the ground.

FIGURE 3.10   Giantess throws the ball around the Earth.

A complication of the real Universe is this: We talk about a planet 'orbiting the Sun,' but the Sun orbits the planet, too. Just as the planet moves under the grasp of the Sun's gravity, the Sun also is required to move under the influence of the planets. The planet performs most of the work in its dance with the Sun. The Sun wobbles almost imperceptibly as the planet whirls around it. Both move around a point between them known as their common center of mass, or barycenter.

The center of mass is not half-way between the two orbiting bodies. Instead, the distance of each from the barycenter is inversely proportional to its mass; the heavier body is closer. Planets have so little mass compared to the Sun that their barycenters with the Sun are actually *inside* the solar radius. The one exception is Jupiter, so massive and so far away from the Sun, that the Sun-Jupiter barycenter is outside the Sun, though only by 7% of the solar radius (Figure 3.13). Similarly, the Earth and Moon orbit their barycenter, which lies about 1,700 kilometers (1,100 miles) below your feet.

So far, we only have considered two-body problems: a planet and the Sun orbiting their barycenter or a satellite and its planet orbiting their barycenter. In our real Solar System, multiple planets and the Sun orbit their common center of mass, some planets have multiple moons, and regions like the Asteroid and Kuiper Belts are jostling with huge amounts of debris—every piece of which is in its own orbit. As each of these bodies approach one another, their mutual gravitational attraction tugs them together. Planetary orbits are never truly elliptical, as stated by Kepler; they are ever-changing due to these never-ending interactions with all the other planets, moons, and similar objects.

FIGURE 3.11 First launch of SpaceX's Crew Dragon spacecraft carrying two astronauts to the International Space Station on 30 May 2020, heralding a new era of space exploration.

Some models of solar-system formation include the ejection of one or more planetesimals. In one such scenario, as a planetesimal came near Jupiter, its gravitational tugging on the massive proto-planet slowed Jupiter's motion and changed its orbit slightly. Simultaneously, Jupiter's gravitational attraction pulled equally on the lesser planetesimal, accelerating it more strongly and sending it careening away. Astronomers have detected objects several times more massive than Jupiter roaming a distant cluster of stars without being gravitationally bound to any particular star.

Space-mission planners use similar interactions, or gravity assists, to reduce the amount of fuel required to propel their spacecraft. NASA launched New Horizons to the Kuiper Belt in 2006 so that the spacecraft could get a boost from Jupiter in 2007, allowing it to reach Pluto in 2015. Although such maneuvers frequently speed up a spacecraft, they also can slow one. MERcury Surface, Space ENviroment, GEochemistry, and Ranging [MESSENGER] passed Earth and Venus, and then it flew-by Mercury three times to reduce its speed relative to Mercury so that it could start orbiting the planet in 2011.

While a single dramatic encounter can drive a planetesimal out of the Solar System or adjust the velocity of a spacecraft, smaller tugs are more common now. Consider playing with a child on a swing; giving him a hard shove out of the swing altogether would be mean. Teasing the child by pushing randomly, sometimes with the motion of the swing and sometimes against it, will have little overall effect. However, pushing at the right moment each time will boost the swing and your four-year-old's enjoyment.

Similarly, if a planetesimal experiences multiple small gravitational interactions in different directions as it moves around the Sun, the pulls will on average counteract one another, leaving its orbit unchanged. If the pulls regularly occur at the same place in its orbit, their cumulative effect

FIGURE 3.12   Fund-raising 'gravity wells' can replicate orbital motion.

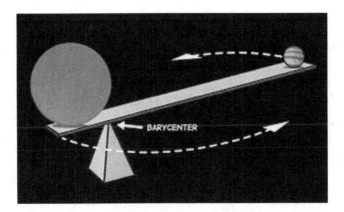

FIGURE 3.13   The barycenter of the Sun and planet Jupiter.

can be either destabilizing or stabilizing. Such aggregate effects are most notable when the two bodies have orbital periods that have simple relationships, known as resonant orbits.

In a destabilizing case, the planetesimal orbit changes, ending the repeated interactions and eliminating the resonance. Gaps occur in the Asteroid Belt where an asteroid would encounter Jupiter once every two or three orbits, which represent 2:1 and 3:1 resonances. ☞ Chapter 9 ☞

In a stabilizing case, the on-going interaction keeps the planetesimal from being too perturbed by other objects crossing its orbit. Many bodies in the Kuiper Belt orbit the Sun either twice for every three orbits of Neptune or once for every two orbits of Neptune. Pluto and other objects near the inner edge of the Kuiper Belt experience a 3:2 resonance, while a 2:1 relationship seems to mark the outer edge.

In 1772, French mathematician Joseph-Louis Lagrange ⟨1736–1813⟩ calculated that, when one body orbits another, five points should exist where the gravitational attraction of both is balanced. Of these so-called Lagrange points, the points 60° ahead and behind the planet are the most stable; these two points are designated L4 and L5, respectively. German astronomers Max Wolf ⟨1863–1932⟩ and August Kopff ⟨1882–1960⟩ confirmed his theory in 1906 with their discoveries of an asteroid in each of these points of Jupiter's orbit: Achilles[3] and Patroclus.[4] ☞ Chapter 9 ☞ These asteroids accompany Jupiter in its orbit of the Sun, a 1:1 resonance.

Resonances also can arise between an orbit and an axial rotation. Our Moon moves once around us and turns once around its axis in the same amount of time, a 1:1 spin-orbit resonance.

Well into the 18th century, people assumed that the laws of motion lead to well-behaved orbits, predicting the future state of the Solar System until the end of time. However, during the latter half of that century, the limits to this classical description of nature became clear.

Under some conditions, gravitational interactions can distort an orbit so that it is essentially unpredictable in the future, or chaotic. The orbit of Toutatis,[5] an asteroid, crosses that of the Earth and takes it out to the Asteroid Belt every 4 years. It is also in a 3:1 resonance with Jupiter. Although it poses no threat to the Earth in the next few centuries, astronomers consider it potentially hazardous because they cannot calculate its path accurately in the distant future. Given enough time, all the orbits in the Solar System might be chaotic. Interactions with Mars have changed the eccentricity of the Earth's orbit. Could a collision be in our distant future? Could the Solar System dissipate as one body after another is expelled?

In 1889, the King of Sweden posed a prize question: Is the Solar System stable or unstable? Do the effects of resonances accumulate resulting in the eventual ejection of a body or are they negligible? French mathematician Henri Poincaré ⟨1854–1912⟩ won the prize by showing there are stable and unstable orbits that can be greatly affected by even the tiniest of disturbances.

FIGURE 3.14 Visualizing orbits of the inner Solar System.

This is the Butterfly Effect: the suggestion, for instance, that a butterfly flapping its wings in China might lead to a hurricane in the US. The field of chaotic dynamics is quite general. It describes movements in such diverse fields as meteorology, chemistry, and cardiology.

In the long term, the Solar System is not a predictable clock; it is chaotic over time. What about the overall stability of our Solar System? Models show that planetary ejections or collisions can occur within the life expectancy of our Sun, but we need not worry about anything catastrophic for the next tens of millions of years!

While this dance of planets is more complicated than first suggested by the elegance of Kepler's and Newton's laws, each body moves as gravity directs. Returning to our simplified view, which adequately describes most of what we observe: Planets orbit the Sun, and moons orbit planets, all under gravity's rule. As you watch a planet move across our sky, remember that it is constantly falling toward the Sun. It just will never quite make it (Figure 3.14).

NOTES

1 A cube has six faces, each of which is a square, while a dodecahedron has 12 faces, each of which is a pentagon. Three other such three-dimensional shapes are possible.
2 Named by South American gauchos.
3 Greek hero of the Trojan War on whom the *Iliad* focuses.
4 Greek hero of the Trojan War.
5 Protector of the Celtic people.

4 Luna, Our Moon

The Greeks called it Selene, the Romans Luna. The second brightest object in the heavens, Greek philosopher Heraclitus ⟨*circa* 500 BCE⟩ explained, was a bowl of fire in the sky. As William Shakespeare's Juliet bemoaned, it is "the Moon, the inconstant Moon,/That monthly changes in her circled orb." Rather than treacherous and inconstant, the Moon has marked time[1] for generations while exerting a stabilizing influence on its host world.

Although the Moon is only slightly smaller than any planet, astronomers do not classify it as one because it orbits a planet, the Earth, rather than the Sun. Yet, we know more about the Moon than any other member of the Solar System save our own world: It is so close and also the only solar-system body beside the Earth that humans have visited (Figure 4.1).

The Moon is the Earth's only natural satellite, the fifth largest in the Solar System. Inasmuch as the lunar diameter is 27% that of the Earth, the Earth and Moon are sometimes called the only double planet in the Sun's family (Figure 4.2).

The Moon appears to rise and set in our sky every day, just like the stars. The reason is the same: The Earth rotates. However, the much closer Moon is not 'fixed' compared to the background stars. The Moon's path appears to encircle the Earth roughly once each month. As it does, its position against the starry background constantly changes. In only a few hours' time, the Moon appears closer to some stars and farther from others than it did when you first began observing that night. It moves eastward about 12° every 24 hours. The lunar motion is the reason that its rising and setting times are about 50 minutes later each day.

The Moon's apparent path against the vault of stars is tipped such that half of it is in our northern sky, and half is in our southern. On those days when the Moon is farther north, it will rise above the horizon in the northeast. On those days when the Moon is farther south, it will rise above the horizon in the southeast. Similarly, it will set in the northwest and southwest, respectively.

When the Moon rises and sets as far north as possible, we say that it is at its northernmost standstill. When the Moon rises and sets as far south as possible, we say that it is at its southernmost standstill.

The word 'standstill' deserves some explaining. After all, the Moon never just stops! Picture a basketball shot at the hoop. As it arcs over toward the net, there is a moment when it is no longer getting closer to the ceiling. A moment later it is getting closer to the floor.

This is what we mean by a lunar standstill: The Moon appears to change directions. Night after night, it rises or sets more-and-more southerly. Then, it reaches its standstill. It will thereafter rise or set more-and-more northerly. The Moon will do this until it reaches a second standstill, at which time it will begin to rise more-and-more southerly again. Think of the direction of moonrise as a swinging pendulum (Figure 4.3).

Since pre-history, people have looked up at the disk of the Moon and wondered about its pock-marked face. Although from the Earth we can only see one hemisphere of the Moon, photographs sent back to the Earth from space reveal the other hemisphere to be similar.

Many people assume that the Moon only shows one side to us on the Earth because it is not rotating. It *must* be rotating, though, to see what we see. Consider the following three situations.

Experiment 1: Turn 360° in place so that you have rotated once. How do you know? You have, in turn, seen each of the four walls making up the room in which you are standing.

Experiment 2: Now, do this while at the same time orbiting, or walking around, another person. Your friend will only see your face. Did you really rotate? Yes. Beyond the person you are facing, you saw the same four walls as you did in the earlier exercise. Perhaps unconsciously, you rotated once in exactly the same time you orbited once. Your rotation and orbital periods were the same.

FIGURE 4.1   Moon as photographed by the crew of Apollo 11.

FIGURE 4.2   Illustrating the relative sizes of our Earth and Moon.

Experiment 3: Finally, make a point not to rotate. Fix your gaze at one point in the room while you walk around your companion. This time, your friend will report that your entire head became visible, face and back, during the orbit.

Over time, we see a single, monotonous view of the Moon rather than a cycle through its vistas. Therefore, the Moon must move in a manner analogous to Experiment 2. To see only one side, the rotation and orbital periods of the Moon must be synchronous, each with a period of one month.

Talking about the nearside of the Moon, the hemisphere that faces us, is appropriate. Similarly, talking about the farside of the Moon, the hemisphere that we cannot see and that is a little bit

FIGURE 4.3   Changing position of moonset. In this mosaic, moonset is shown for two different days of the month (left and right). For reference, sunset is shown in the middle.

farther way, is appropriate. Nevertheless, speaking of a 'dark side' of the Moon is nonsense.[2] Because the Moon rotates, both the farside and nearside experience day and night, with each lasting about two weeks from sunrise to sunset and sunset to sunrise.

Although we can periodically peak ever-so-slightly at a portion of the farside, we knew almost nothing about that region until the Soviet Luna 3 orbited Moon in 1959. The US Lunar Orbiters eventually followed in 1966–1967 (Figure 4.4).

When the Sun totally illuminates the lunar nearside, the Moon certainly looks bright—it is the night of the Full Moon. Admittedly, though, this perception is partially due to the absence of another bright disk in the sky (the Sun) with which to compete. At this time, the Moon appears directly opposite the Sun in the sky; it rises at sunset and sets at sunrise (Figure 4.5).

The Full Moon looks huge as it rises or sets. This is not because the Moon is necessary closer to the observer. The apparent size difference between the Moon when it is closest to the Earth (perigee)[3] and when it is farthest (apogee) is small because the eccentricity of its orbit is low.

FIGURE 4.4   Lunar farside as imaged by the Lunar Reconnaissance Orbiter [LRO]. Compare to Figure 4.1 showing the nearside.

FIGURE 4.5 Full Moon rises.

Instead, this image is a psychological effect. Only at the horizon do we have objects (trees, buildings, *etc.*) with which to compare the Moon, making it look relatively close and our brain perceives its size to be larger. These comparative objects are absent when the Full Moon is high in the sky (Figure 4.6).

Traditionally, Full Moons have names, taken from Native-American sources. For instance, the Harvest Moon is the Full Moon that takes place at the beginning of autumn. In northern climates, this is often the time to bring in the crops, a time-consuming process. A little extra light after sunset is appreciated. Following the Harvest Moon is the Hunter's Moon.

Over the next few nights, the Moon will appear to rise a little later and will appear a little less full. It is moving through its waning gibbous phase. A waning body is weakening in appearance. From the Latin for 'humped,' gibbous describes a celestial body that is more than half illuminated or that has an additional lit 'hump' protruding from its illuminated half. The gibbous Moon looks odd, even though the Moon is gibbous for the same amount of time as it is crescent. (See below.) Artists who portray the Moon often use a crescent in the sky (or a Full Moon if they are going for a spooky effect). However, you rarely see a gibbous Moon in art.

About a week after Full Moon, only half of the nearside of the Moon is illuminated. If you live in the North, the illuminated eastern half is to your left, and vice versa in the South. This Third Quarter Moon rises around midnight, is visible shortly after dawn, and sets about noon. While there is an

FIGURE 4.6   Depicting the apparent sizes of the Moon at perigee and apogee.

instant when the angle between the Sun, Earth, and Moon is 90°, in practice there is an entire night when the line dividing the day and night sides, or terminator, on the Moon looks straight. Quarter Moon? Clearly half of the lunar disk is visible; this term refers to the fact that about three weeks have passed since the traditional start of the month: the date is three-quarters of the way through the month. Although sometimes called Last Quarter, it is definitely not 'Half Moon'!

For the next week, moonrise will continue to move through the early morning hours while the illuminated section of the Moon dwindles to a crescent. This waning crescent phase lasts until the Moon 'disappears' at New Moon, at which time its farside will be fully lit but our familiar nearside will be in shadow. The New Moon is closely aligned with the Sun in the sky so that it rises at sunrise and sets at sunset lost in the bright sunlight. Note that the New Moon does not occur when the Moon is first visible at the beginning of the month: It is when one cannot see the Moon all!

About a day after New Moon, dedicated observers may glimpse the first waxing crescent. A waxing body appears to strengthen. Over the next week, the nearside of the Moon will be increasingly illuminated. It will rise later in the morning and set later in the evening, making it easier to see after dusk.

When the lunar nearside is again half illuminated, the Moon will have reached its First Quarter, which rises around noon and sets near midnight. The angle between the Sun, Earth, and Moon is again 90°, on the other side of the Moon's orbit. At First Quarter, the opposite half-disk of the nearside (western half) is lit as compared to Third Quarter. Next, a 'hump' will appear out of the illuminated half that will swell throughout the waxing gibbous phase as moonrise occurs later in the afternoon. After one lunation, the time it takes the Moon to go from one phase to that same phase again, the Moon will again be Full[4] (Figure 4.7).

The interval of time the Moon takes to go through a cycle of lunar phases is called the synodic month (29.5 days). It is slightly longer than the time the Moon takes to orbit the Earth as measured against the background of stars, the sidereal month (27.3 days). We pay more attention to the synodic month than the sidereal because the changing relative positions of the Earth, Moon, and Sun enable us to see different phases.

Moon phase cycle, from 1 day
to 28 days old (totally dark)

FIGURE 4.7   Phases of the Moon.

Think of running a one-lap foot race on a circular track. During the race, somebody picks up and moves the start/finish-line marker farther down the track. You must run a greater distance to complete the event, traveling more than 360° around the track; your time will also be longer.

The Sun is the start/finish marker for the synodic month. As the Moon moves around the Earth, the Earth moves around the Sun. After a sidereal month, the Moon has traveled once around the Earth (360°) and the Earth has shifted about 1/13 (27°) of its way around the Sun. Therefore, the three bodies are not quite aligned in the same phase as they had been 27.3 days ago; the marker appears to have moved. Over the next two days, the Moon continues to move around the Earth, and the Earth continues to move around the Sun. Then, the three will re-align in the appropriate phase. Thus, the Moon must travel more than 360° to complete a cycle of phases. The time needed to advance this extra distance results in a synodic period longer than the sidereal one.

Occasionally, the Moon intersects the shadow cast by the Earth and undergoes a total lunar eclipse. Such an event does not occur every month because the plane of the Moon's orbit around the Earth and the plane of the ecliptic are offset by 5.1°. The two planes intersect at points called nodes. The Moon is at one of its nodes (otherwise empty points in space) twice a month: the ascending node and the descending node.

The Moon's orbit precesses (wobbles) around the plane of the ecliptic in the direction opposite the Moon's motion. So, the time it takes to go from node to node to node is different from a synodic month.

Full Moon must occur while the Moon is at a node for a total eclipse of the Moon to take place. This happens approximately twice every three years (Figure 4.8).

During a total lunar eclipse, the Moon passes completely through the inner core of the Earth's shadow, the umbra. For several hours, everyone on the night side of the Earth will see a dark, coppery disk. The Moon does not disappear altogether because some sunlight bent through the

FIGURE 4.8 2018 total eclipse of the Moon.

Earth's atmosphere at the terminator still reaches it. As light travels through terrestrial air and dust, the blue colors are scattered away leaving the red, similar to a sunset.[5]

Partial and penumbral lunar eclipses are also possible, but you will not detect either easily with your eyes alone. A partial lunar eclipse occurs when only a fraction of the lunar disk passes into the Earth's shadow; a penumbral lunar eclipse occurs when the Moon only enters the outer edge, or eclipse, of the Earth's shadow. As seen from any place on the lunar nearside, the entire Sun is blocked during a total lunar eclipse, but part of the Sun remains visible during a partial lunar eclipse.

At the opposite side of its orbit, the Moon occasionally passes directly in front of the Sun during a New Moon while at a node. Again, the inclination of the lunar orbit to the Earth's orbit prevents this from happening every month. However, when these three bodies align just so, a small swatch of the Earth will briefly see a total solar eclipse with the Moon completely blocking the Sun. Because the shadow of the Moon is so small compared to the surface of the Earth, a total eclipse of the Sun is visible from only a tiny fraction of the Earth's surface (Figure 4.9).

For example, the 21 August 2017 total eclipse of the Sun across the continental US was visible along a 120-kilometer (70-mile) wide path (at maximum) from Oregon to South Carolina for no more than 2 minutes, 40 seconds. As seen from most of the rest of North America, the Moon only partially obscured the Sun; these people experienced a partial solar eclipse. Totality can range from

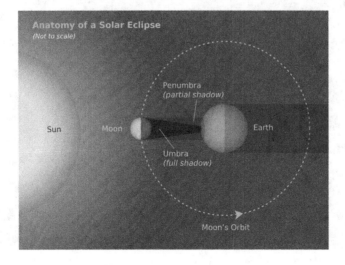

FIGURE 4.9 Geometry of a solar eclipse.

a fraction of a second to a maximum of 7½ minutes (Figure 4.10). In every case, the Sun spends several hours in partial phases as it enters and then exits the shadow. VIEWING A PARTIAL SOLAR ECLIPSE SAFELY REQUIRES APPROPRIATE EYE PROTECTION.

While a total solar eclipse may be visible somewhere on Earth about every 18 months, a particular location may go hundreds of years between events. Unusually, another total eclipse of the Sun will be visible across the US, this time from Texas to Maine, on 8 April 2024. Carbondale, Illinois, is where the 2017 and 2024 total-solar-eclipse paths cross. This city will only have to wait 7 years between total eclipses (Figure 4.11).

On 14 October 2023, people along a path from Oregon to Texas and further into South America will experience a particular kind of partial solar eclipse. In this case, the Moon will be a little farther away from the Earth in its orbit than average so it will appear smaller than usual. Therefore, it will not cover the Sun completely but will allow a ring, or annulus, of sunlight to remain visible (Figure 4.12). An annular solar eclipse like this can occur either when the Earth is near perihelion or the Moon is near apogee. Please do not confuse annular (ring) with annual (year)!

Eclipses of the Sun are possible because of a coincidence: The Moon and Sun take up the same amount of sky—they both are circles ½° in diameter, about the size of your 'pinky' at arm's length. The Sun is 400 times bigger than the Moon, but the Moon is 400 times closer! (The Sun may look bigger in isolation, but that is likely because it is glaringly bright.) No other planet has a satellite of the right physical size or distance to cover the Sun as seen at the planet's distance. What luck it is for the sole sentient inhabitants of the Solar System to live at the right place to witness this spectacle!

The lunar surface is covered with craters, big and small. 'Crater' means 'cup' in Greek, and, indeed, these round features are depressions in the ground. The globe of the Moon is very different from that of the Earth. It is populated with these landforms that seem alien. However, many solar-system bodies have similar landscapes; *Earth* is an exception.

FIGURE 4.10   2019 total eclipse of the Sun.

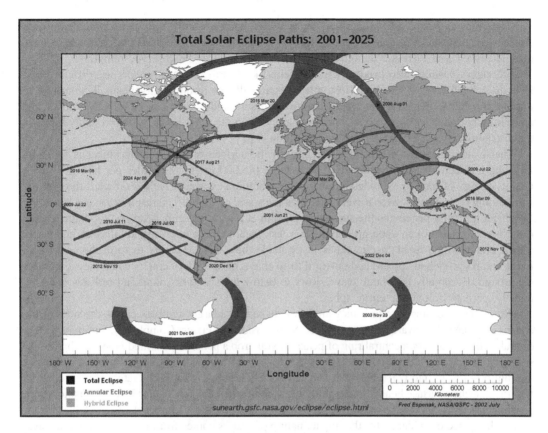

FIGURE 4.11   Total-solar-eclipse paths, 2001–2025.

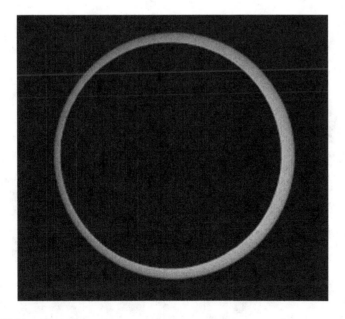

FIGURE 4.12   2014 annular eclipse.

Meteoroids and asteroids, small rocky and metallic bodies orbiting the Sun, form an explosion crater when they strike a solar-system surface. ☞ Chapter 9 ☞ Some 500,000 lunar craters have been identified, and there are doubtless many more too small to be observed remotely.

Although volcanoes also produce circular, cup-like features, impact craters are differentiated from those by their cross sections. Most volcanoes are tall with a comparatively small depression at the top, the caldera. Impact craters are shallow and surrounded by a rim of hills. Moreover, the apparent distribution in size of crater-producing bodies in the inner Solar System roughly matches the distribution of crater sizes found on the Moon.

What exactly happens when an object strikes the surface of Moon? The object is typically traveling between 20 and 72 kilometers per second (45,000 to 160,000 miles per hour). When it hits, its energy of motion explodes into heat and vibrational waves traveling outward through the surrounding rock. These waves travel faster than the speed of sound. The result is a shock wave. As the object normally penetrates the surface to a depth at most a few times the body's diameter, the result is reminiscent of an underground nuclear test. Debris flies away, only to fall back and partly fill the crater. Much rubble accumulates just outside the exploded cavity to form the crater rim. Large chunks propelled upward and outward form chains of secondary craters when they, too, meet the ground. Eventually the shock wave slows to below the speed of sound and radiates out seismically, perhaps around the entire sphere.

Most craters are simple craters; they are bowl-shaped. In a larger complex crater, the sides of the crater rim slump downward leaving terraced walls inside the crater. Complex craters may have a central peak or peaks, or a central ring of peaks. If you drop a pebble into a mud hole, you had better step back so that you are not covered with the recoil 'blurp' of mud! The central peak of a crater is the 'blurp'—frozen into place before it can fall back. The objects forming craters are traveling so fast, the planetary surface behaves essentially like a fluid when they encounter it. The transition from a simple crater to more complex forms varies with crater diameter (Figures 4.13 and 4.14)

Another piece of evidence for the impact nature of craters comes from features, known as rays, which radiate from some craters. Geologists think that fresh material from just below the surface

FIGURE 4.13 Simple crater imaged by LRO.

FIGURE 4.14   Central peak in the complex crater Tycho, imaged by LRO.

forms these superficial markings. This material spreads out for hundreds of kilometers at impact, just like a lump of flour adopts a rayed appearance when dropped on the kitchen counter. Rays never show any relief. With few sources of erosion on the Moon, these delicate streaks remain visible for a long time, highlighting the site of an impact crater (Figure 4.15).

Really big impact craters are called impact basins. The Moon boasts some of the largest such features in the Solar System. The collision of a 200-kilometer (120-mile) wide asteroid and the Moon resulted in Aiken Basin, near the Moon's South Pole. It is 2,500 kilometers (1,600 miles) in diameter and 13 kilometers (8 miles) deep—the lowest elevation on the Moon. In 2019, the Chinese space probe Chang'e[6] 4 landed there (Figure 4.16).

FIGURE 4.15   Crater with rays photographed by the Apollo 13 crew.

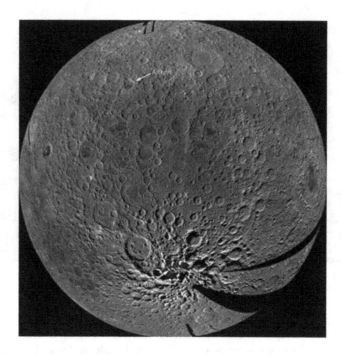

FIGURE 4.16   Aitken impact basin imaged by LRO

In reality, the rough lunar surface is a pretty *poor* reflector.[7] Recall that its *average* albedo is only 0.11. But the large round features that make the face of the 'Man in the Moon' have a lower albedo and appear darker than their surroundings. These lower-albedo regions are named *maria*, which is Latin for 'seas,' because of the one-time mistaken belief that they were smooth bodies of water. Maria cover 16% of the lunar surface. As orbiters pass over the maria, they speed up slightly, indicating mass concentrations beneath, named mascons. Higher-albedo and heavily cratered lunar highlands surround the darker maria. Puzzlingly, there are more maria and less highlands on the nearside than on the farside.

If you encountered a divided highway with one smooth lane and one lane covered in potholes, you would assume the smooth side had a new surface due to recent roadwork. Similarly, a lunar region with few craters is considered relatively 'young.'

Imagine a Moon uniformly cratered. Then, one day, a large object crashes into it, releasing dark molten rock (lava) that 'repaves' the site—now a mare. Long thereafter, the crater density in the mare remains less than that of its surroundings. By counting craters and assuming fewer craters implies a younger surface, we can infer the relative history of the Moon: random cratering punctuated by a few large impacts. Close-up dating further relies on a simple rule of superposition: A crater overlapping another is the younger of the two.

To use this technique, planetary geologists must account for a changing cratering rate. It was particularly intense early in the history of the Solar System, during the Era of Heavy Bombardment. ➤ Chapter 2 ➤ Since, cratering has trailed off. This places the highlands as the oldest regions on the Moon compared to much younger maria.

Radioactive elements within rocks will slowly change by losing subatomic particles. The amount of such radioactive decay a rock has experienced gives its absolute age. A single rock sample returned from the Moon, combined with the tool of superposition described above, will yield the age of the swath on the lunar surface from which the rock came.

The result of these two dating techniques? The battered highlands were formed about 3.9 billion years ago and the maria between 3.9 to 3.2 billion years ago. Most of the lunar features we see today have been there for more than 3 billion years!

The Moon has little obliquity. For some lunar craters near the poles, their bottoms are permanently in shadow because the Sun never rises to a high altitude in the sky above them. Here, it is always cold, and any water that may make it to the Moon could survive in a frozen form almost indefinitely. First hinted at by 1991 radar analysis, imaging of the Moon shows that this really is happening: Pools of perpetual darkness hide old ice—otherwise unexpected on this dry world.[8] So, if you intend to visit the Moon, a ready supply of water may be one less thing you must carry on your back.

All the Apollo[9] expeditions ⟨1969–1972⟩ landed far from the poles, on the nearside; most touched down in a mare. They found the ground covered by a blanket of loose, grayish-charcoal rock fragments and dust created by much earlier impacts; such unconsolidated rock is called regolith. The result is the iconic photograph of a boot print on the Moon (Figure 4.17).

With estimated depths of up to 120 meters (39 feet) of dust, some scientists worried that our astronauts and their spacecraft might disappear upon touchdown! The successful soft landing of Surveyor 1 on 2 June 1966 led the way, proving that this would not be the case (Figure 4.18). In fact, astronauts described the surface layer as quite solid seeming. More troubling, the fine abrasive rock got into everything, wearing down spacesuits and other equipment. Without an atmosphere, no lunar breeze blows away the dust. The dusty, regolithic surface the astronauts walked upon is thought to give way at a depth of 2 kilometers (1 mile) to more compressed, larger pieces of rock called the megaregolith.

The Apollonauts planted the US flag. In this silent, airless environment, no weather stirs their banner, which is displayed against the black sky. They also drilled into the Moon, picked up sample rocks, and produced a great number of photographs. An ultraviolet camera, invented by African-American astronomer George Carruthers ⟨1939–2020⟩, was used by astronauts on Apollo 16 to image astronomical objects from the lunar surface, opening the modern era of ultraviolet astronomy. A total of six missions landed on the Moon, carrying 12 humans who spent more than 80 hours on the lunar surface. NASA planned even more ambitious landings, perhaps in the highlands, in one of those centrally peaked craters, or on the lunar farside. However, they cut these missions in the face of what was perceived as waning interest back home—and the need to pay for the Vietnam War (Figure 4.19).

So, the Apollo missions were much like loading the family into the minivan, driving 1,600 kilometers (1,000 miles) to Yellowstone National Park, and then sitting in the car park for 20 minutes. Upon

FIGURE 4.17   Boot print left by the Apollo 11 crew.

FIGURE 4.18   Apollo 12 astronaut retrieves part of Surveyor 3.

FIGURE 4.19   Apollo astronaut saluting on the Moon.

returning, you tell everybody that you have seen Yellowstone! Clearly, the Moon has a lot more to tell us (Figure 4.20).

On the bright side, geologists still have 380 kilograms (840 pounds) of rock samples plus other data collected. These can be subjected to experiments that were not even thought of back in the 20th century. We are still learning about the Moon (Figure 4.21).

For instance, rocks from the lunar surface are characteristically older than those from the Earth, as determined by radioactive dating techniques. The abundances of some lunar rock-bearing minerals are the same as on the Earth, others have different abundances. Moonrock from the maria is basaltic, like Earth's ocean beds. Moonrock from the highlands consists of the mineral anorthosite

FIGURE 4.20   Apollo astronaut examines a lunar boulder. It may have been excavated from within the nearby crater.

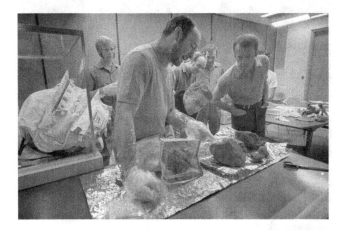

FIGURE 4.21   Processing the Apollo 14 moonrocks.

and breccias (bits of other rocks compressed together). Nowhere are there appreciable volatile elements (Figure 4.22).

The fact that moonrocks are like terrestrial rocks in some ways and quite different in others, led to a new theory of the Moon's origin. Previously, scientists speculated that the Moon formed in place next to its larger neighbor, was spun off a more rapidly rotating Earth, or was gravitationally captured whole when it approached the Earth from elsewhere.

The first theory is plausible, but nothing like it took place anywhere else in the inner Solar System. The 'spun-off' theory was motivated by the coincidental sizes of the Moon and the Pacific Ocean Basin. Was the Pacific Ocean the scar from the Moon breaking away? However, according to the theory of plate tectonics ☛ Chapter 7 ☛, the Pacific did not exist at the time of the Moon's formation. The third theory relies on a very serendipitous trajectory for the invading body: It would have had to approach from just the right direction, at just the right speed, and somehow lose enough energy to orbit. It is much easier to collide with the Earth than to enter a nearly circular orbit around it.

Regardless, if the first or second theory were true, the moonrocks should be chemically identical to those on the Earth. If the third theory were true, the two types of rocks should be completely different. None of these theories explain the actual composition.

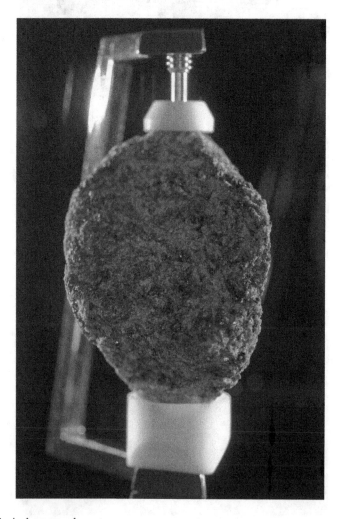

FIGURE 4.22   Typical moonrock.

Scientists now think that a body the size of Mars, or greater, struck the Earth a glancing blow, ejecting a mix of excavated terrestrial material *and* material that formed elsewhere in the Solar System into orbit about the Earth. For a while, the Earth was encircled by a ring. This debris later coalesced to form the Moon. The event may have taken place early—within the first 100 million years since the formation of the Solar System. What a sight it would have been—from a safe distance! The explosive origin of the Moon would have vaporized any volatile elements thrown into orbit. These would have escaped, leaving the volatile-poor Moon we see today (Figure 4.23).

Collecting rocks only scratched the lunar surface. Four Apollo missions installed seismometers, instruments to measure tremors moving through the Moon. Using data collected between 1969 and 1977, planetary geologists deduced the structural composition of the Moon by mapping where these vibrations pass through it and where they are blocked or deflected.

The Moon is geologically inactive; for instance, volcanoes ceased to erupt there long, long ago. Still, moonquakes do occur occasionally. There are three kinds: those generated 1,000 kilometers (600 miles) below the surface due to tidal effects, those triggered at the surface by impacts and diurnal thermal expansion/contraction, and those that take place 30 kilometers (19 miles) below the surface due to yet unknown causes.

The result of lunar seismic and gravity studies is: (1) a high-density, partially molten region extending from the radius of the Moon's inner core to about 700 kilometers (430 miles) from the center of the Moon with an iron-rich inner core only 240 kilometers (150 miles) in radius; (2) an outer core, extending to about 330 kilometers (210 miles) from the center of the Moon; (3) a mantle shell, of solid high-density rock, extending to the boundary with the crust; and (4) a thin outer crust of lower-density rock averaging 50 kilometers (30 miles) thick. Deep moonquakes originate at the boundary of the solid mantle and the partially molten zone. Neither core nor mantle generate a magnetic field.

The lunar core extends to only 20% of the Moon's total radius. This is small: The differentiated (layered) terrestrial planets have cores that reach at least halfway out from their center. The Moon's core compromises only 2% of the Moon by mass. Interestingly, its center, the center of mass of the Moon, is displaced 2 kilometers (1.2 miles) from the geometric mid-point of the satellite in the direction of the Earth. The cause is unknown.

The lunar crust deviates in thickness from the average to essentially none in the bottom of Mare Crisium or to 110 kilometers (68 miles) thick north of the Korolev Plain. For some reason, the crust is deeper on the lunar farside than it is on the nearside (Figure 4.24).

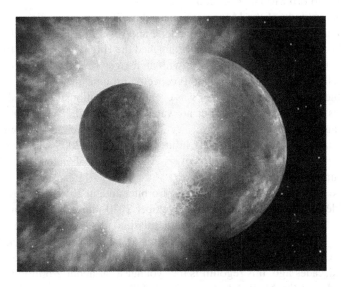

FIGURE 4.23   Artist's rendering of how the Moon formed.

FIGURE 4.24 Moon in cross section.

This lunar structure is consistent with our impact model for the formation of the Moon. The dense core of the foreign body would have merged with that of the nascent, liquid Earth. It would have been the lower-density elements, such as those that constitute rock that would have splattered disproportionately into orbit to make a Moon.

Meanwhile, remote study of the Moon continues. Now, it is an international affair. In addition to the US and Soviet lunar probes, orbiters, landers, and rovers of the 1950s through 1970s, the Japanese joined this exclusive group when they launched the Hiten[10] mission to the Moon in 1990, using a highly elliptical orbit around the Earth with an apogee near the Moon. In 1994 and 1998, respectively, the US Clementine and Lunar Prospector orbiters mapped the lunar surface and discovered ice at the poles. Lunar Prospector was eventually intentionally crashed into the Moon to see what chemicals might be excavated, but no water was detected. The Japanese Kaguya[11] was launched in 2007 and orbited the Moon for nearly 2 years, after which it was commanded to crash near a lunar crater in 2009. India successfully orbited the Moon in 2008 with its Chandrayaan-1.[12] In 2011, the two-orbiter NASA's Gravity Recovery And Interior Laboratory [GRAIL] mission was sent to study the lunar gravitational field.

China joined the club in 2007 with the Chang'e 1 orbiter and, in 2019, with rovers, one of which explored the farside for the first time. On 23 November 2020, the China National Space Agency launched Chang'e 5, an ambitious sample-return mission to Oceanus Procellarum (Ocean of Storms). Less than 4 weeks later, almost 2 kilograms (4 pounds) of moonrock landed in Inner Mongolia. Compared to samples obtained by the USA and the Union of Soviet Socialist Republics [USSR] collected decades ago, this regolith is the 'youngest' at about 2 billion years old, filling a gap in our knowledge of the Moon's history (Figure 4.25).

FIGURE 4.25 China's Yutu 2 Rover deploys on the Moon.

Undoubtedly, humans will return to explore the Moon and possibly establish a colony, exploiting lunar resources, and gaining important skills for a maiden voyage to Mars. Whose will be the first country to do so? A new 'space race' has begun! ☞ Chapter 17 ☞

NOTES

1 The word 'month' comes from 'moon.' The lunar cycle was the basis of many ancient calendars, some still in use today.
2 It is, though, the title of a classic Pink Floyd album. They made great music but had dubious astronomical understanding.
3 From Ge, an alternative form of Gaia, the Greek goddess personifying Earth.
4 While the word 'lunatic' comes from Luna based on the superstition that the Full Moon drives people insane, hospital and police records do not indicate an upsurge of strange behavior during this phase.
5 For this same reason, our sky is blue.
6 Named after the Chinese Moon goddess.
7 Only because of the bright sunlight pouring on it can the Moon be seen in the daytime.
8 Have you heard the expression, "where the Sun doesn't shine"? It is definitely referring to these permanently shadowed regions on the Moon!
9 Greek god of prophecy and music who is also associated with the Sun.
10 Which means 'Flying Angel.'
11 Which means 'Moon Princess.' JAXA renamed the mission after launch; it was previously SELenological and ENgineering Explorer [SELENE].
12 Which means 'Moon-craft.'

5 Sun-scorched Mercury

We will not spend a *lot* of time with the first planet from the Sun. We will use Mercury as an example of the kind of things found on other terrestrial planets. In doing so, you will notice that many of our expectations for how a planet 'should' behave—based on our familiarity with the Earth—must be left at the proverbial door. On the other hand, you will see many similarities between Mercury and the Moon.

Mercury is hard to see. The smallest planet, it is but 4,900 kilometers (3,000 miles) in diameter, which is a little more than a third of the Earth's diameter and barely larger than the Moon. Indeed, a moon of Jupiter and one of Saturn are both bigger than Mercury, although Mercury is more massive than those. Its orbit is the most elliptical of all planets, only 0.31 au from the Sun at its perihelion and 0.47 au at its aphelion. Like Venus, Mercury is an inferior planet, being closer to the Sun than the Earth, and moves along an inner orbit. This means that it is never more than 28° away from that bright orb in the sky, so it gets lost in the glare much of the time. Most people have never seen this elusive planet.

Consequently, early observers thought Mercury was *two* planets: a lesser morning 'star' when it appeared in the Eastern sky just before dawn and an evening 'star' when it lingered in the Western sky after dusk; they made a similar mistake with Venus. ☛ Chapter 6 ☛ Nonetheless, a compilation of Babylonian astronomical records, known as the *Mul Apin*, reports observations of it dating to *circa* 1400 BCE. They referred to it as 'jumping' and called it Nabu, which is the scribe of their chief god, Marduk. The Romans associated Nabu, the patron of writers, with their god Mercury,[1] the fleet-footed and glib messenger of the Olympians. In the Greco-Roman tales, Mercury demonstrated his cleverness through his inventions, such as musical instruments, and the occasional theft.

Periodically, Mercury crosses the disk of the Sun. ☛ Chapter 2 ☛ Transits of Mercury occur 13–14 times a century, which is more often than those of Venus ☛ Chapter 6 ☛ but less frequent than solar eclipses. After the 11 November 2019 transit of Mercury, we will have to wait until 13 November 2032 for another (Figure 5.1).

Mercury revolves about the Sun in a mere 88 of our days, which constitutes its year. So, its position in the sky is always rapidly changing and easy to lose.

Mercury not only has a year, but it also has a day. Like all the planets, it rotates about its axis. For many years, astronomers thought its rotation period was the same as its orbital period; after all, the Moon takes a month to turn on its axis and a month to orbit the Earth. In the case of Mercury, that would mean that one side always would be in sunlight and the other hemisphere would be in perpetual night.

Radar, however, revealed that Mercury was different. In 1965, measurements at the Arecibo Ionospheric Observatory[2] revealed that Mercury takes 59 of our days to complete one slow rotation.

On Earth, the day and year are very different units of time; a day is only 1/365 of a year. At the other extreme, the ratio of the Moon's rotation and orbital periods is 1:1. Yet, on Mercury, that ratio is 2:3. The coupling of the rotational and orbital periods is an example of a resonance. ☚ Chapter 3 ☚ It is the only planet with a spin-orbit resonance. We shall see more examples of orbital resonances and their dramatic effects in later chapters.

This 2:3 ratio also means that two opposing regions near Mercury's equator alternatively line up under the intense mid-day Sun every two revolutions, resulting in planetary hot spots. Ninety degrees in longitude from the hot spots are equatorial areas that are relatively cooler. Already, individual planets are diverging from Earthly expectations (Figure 5.2).

FIGURE 5.1 Mercury (upper left) transits the Sun's disk.

Why does Mercury take so long to turn about its axis? One theory is that the planet once spun at a much faster rate—perhaps as fast as 8 hours per rotation. Are tidal effects the culprit? ☞ Chapter 7 ☞ Or, more spectacularly, did a tremendous off-set collision long ago affect Mercury so fundamentally that it changed the planet's rotation? Such events would have been more common in the early, more-crowded Solar System. If such an occurrence happened while Mercury was still young and molten, it would have left no visible trace.

Mercury's orbit about the Sun is quite eccentric; that is, it is more elliptical and less circular, and more inclined than other planets. Remember that all planets travel fastest at perihelion and slowest at aphelion. Meanwhile, a planet's rotation rate is nearly constant. These facts combine to produce a remarkable sight from some locations on Mercury: In the morning, an astronaut would watch for the appearance of the Sun. The Sun does rise, but then appears to *reverse* direction and set again—only to rise once more and continue with the day. As the Sun sinks slowly in the west, the astronaut would see the same phenomenon. The cycle repeats itself every 176 Earth days, just over two of Mercury's years.

A calendar on Mercury would be exceedingly simple; starting with January 1st, only a half-day would occur before it was New Year's Day again. (Would New Year's Day be a holiday?) Babies

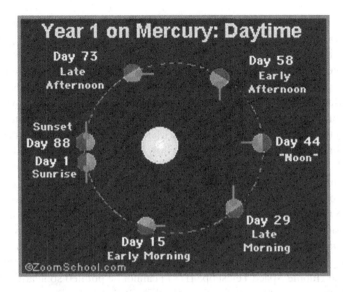

FIGURE 5.2 Rotation and Revolution of Mercury.

born there on the same day would note two different years on their birth certificates, depending if they entered this world in the 'morning' or 'evening.'

At midday, from certain longitudes, the Sun may appear three times as large as it does from the Earth—and it is seven times brighter. Nevertheless, the surrounding sky is black and star-speckled inasmuch as Mercury does not have an atmosphere in the traditional sense. It is a lonely sky, without a satellite nor even rings, orbiting the planet.

Mercury's orbit slowly shifts around the Sun. That is, its point of perihelion moves. This is to be expected due to the overpowering influence of the Sun's gravity. However, Mercury's so-called precession is greater than can be accounted for by the Sun and other planets alone (Figure 5.3).

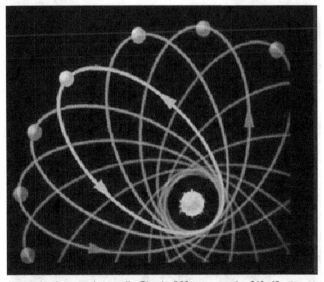

http://physics.ucr.edu/~wudka/Physics7/Notes_www/img249.gif Fig; 43

FIGURE 5.3 Simulation of Mercury's precession (not to scale).

In the 19th century, Urbain Le Verrier ⟨1811–1877⟩, a French astronomer specializing in calculating orbits, postulated a terrestrial planet sunward of Mercury to account gravitationally for the anomalous orbit. As early as 1859, observers reported tentative sightings of this 'Vulcan' transiting across the solar disk. It remained controversial and eventually proved to be spurious: German-American physicist Albert Einstein's ⟨1879–1955⟩ General Theory of Relativity eventually explained Mercury's strange behavior without resorting to a new solar-system body.[3]

Mercury goes through phases; in fact, the planet is best viewed during its quarter phases when its separation from the Sun is greatest. The face of Mercury was first observed through a telescope by Galileo in 1609 and later, more carefully, by fellow Italian Giovanni Zupi ⟨1589–1650⟩. They saw a tiny disk, but its apparent size still was too small for phases to be seen. Nevertheless, Mercury's albedo of 0.12 is only slightly higher than that of the Moon.

Because of its poor position for observation from the Earth, astronomers knew little more about Mercury for a long time. It was sort of like the statistics on the back of a trading card—without the face on the other side.

When the Mariner 10 space probe flew past Mercury three times between March 1974 and March 1975, it began to fill in the details by sending back pictures of about 45% of Mercury's surface. Because the Hubble Space Telescope [HST] cannot be pointed so near the Sun, the further investigation of Mercury had to wait for the MESSENGER robotic spacecraft. It flew past the planet three times in 2008–2009 before settling into orbit in 2011 where it remained, mapping its entire surface, until it ran out of fuel and crashed in 2015. These missions have dramatically increased our knowledge of this solitary world (Figure 5.4).

Images transmitted to the Earth from robotic space missions at Mercury show a surface heavily covered by craters, most likely formed during the Era of Heavy Bombardment. The landscape of Mercury highly resembles the Moon, not the Earth.

Although even trained planetary scientists sometimes have difficulty telling Mercury and the Moon apart, one difference is the featureless intercrater plains of Mercury. Small craters seem to have been mysteriously covered up. Volcanism? The origin of Mercury's intercrater plains remains

FIGURE 5.4 Mercury as imaged by the MESSENGER mission.

unknown. Distinctly different from the intercrater plains, about 40% of Mercury's surface is covered by smooth, volcanic plains that have filled old craters and basins. These resemble lunar maria (Figure 5.5).

Another difference is the appearance of more complicated craters at smaller diameters on Mercury than on the Moon. Crater diameter is a function of gravity, and Mercury is more massive (Figure 5.6).

The largest impact feature on Mercury is Caloris Basin, which is at least 1,500 kilometers (950 miles) across, the size of Texas. It is one of the largest impact basins in the Solar System. Odd hilly terrain covers the spot directly opposite from Caloris. Apparently, the formation of Caloris was so violent that its waves rippled through and around the entire planet and came to a focus to form these complex hills. That would have been a bad day to be standing anywhere on Mercury! (Figure 5.7)

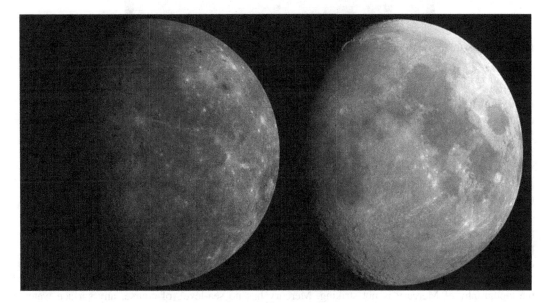

FIGURE 5.5 Mercury (left) and the Moon (right) in similar illumination.

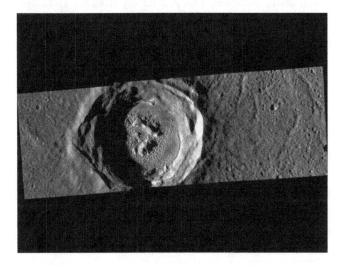

FIGURE 5.6 Complex crater on Mercury as imaged by MESSENGER.

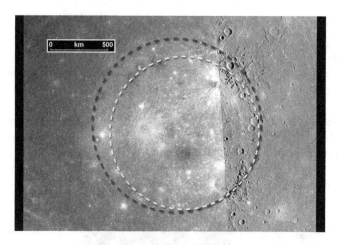

FIGURE 5.7   Caloris Basin as imaged by MESSENGER.

Mercury is also home to dramatic scarps (cliffs). Many are hundreds of kilometers long and more than a kilometer tall. Some look geologically young since they cut across more ancient craters. Two such scarps, Enterprise Rupes[4] and Belgica Rupes,[5] outline the Great Valley of Mercury, which is sunk between them. This valley is 1,000 kilometers (620 miles) in length and 3 kilometers (2 miles) deep. These 'wrinkles' are the result of Mercury shrinking—reducing its volume while its crust 'tries' to maintain the same area. Mercury may be a kilometer smaller in radius than it was originally. Indeed, Mercury is even now shrinking as it continues to cool (Figure 5.8).

Thus, like what we will see on the Earth, and unlike the Moon, Mercury is somewhat geologically active. ☞ Chapter 7 ☞ Geological activity refers to processes that take place within a planet as opposed to those which are inflected from without, *i.e.*, cratering. Mercury even may experience occasional 'Mercury-quakes.'

All that said, in the absence of major geologic activity in the form of plate tectonics ☞ Chapter 7 ☞, the topography of Mercury is not striking. Mercury has no sea-level, of course; any surface water would evaporate in the harsh Sun and escape the planet! Still, astronomers establish a mean radius with respect to which they measure variations. It turns out that the highest point on Mercury is 4.5 kilometers (2.4 miles) above the mean. This is a far cry from the Himalayas on the Earth with

FIGURE 5.8   Scarp on Mercury bisects an older crater in a MESSENGER orbiter image.

multiple peaks above 7.2 kilometers (4.2 miles). The lowest point is in the middle of Rachmaninoff[6] crater at 5.5 kilometers (3.5 miles) below the mean; in comparison, Challenger Deep in the Earth's Mariana Trench descends more than 11.0 kilometers (6.8 miles).

Like the Moon, Mercury has essentially no atmosphere because its low gravity cannot hold onto one. Instead, Mercury has a rarified exosphere, or sparse layer of escaping sodium, potassium, and other trace elements. While a few of these are neutral atoms, most are electrically charged ions that have lost one or more their outer electrons. The solar wind knocks these particles off the planetary surface and blows them away, creating a tail of matter on the side of Mercury opposite the Sun.

The Sun beats continuously and directly upon Mercury's exposed surface. Consequently, the temperature may soar to 430°C (810°F).[7] However, that same surface cools down rapidly after sunset, as heat escapes directly into space. During the long nighttime, the temperature may plummet to −180°C (−300°F).[8] Not unexpectedly, Mercury is one of the hottest places in the Solar System, but it is also one of the *coldest*. It is a place of remarkable contrasts.

Mercury is about the last place you would look for water. As on the Moon, there are craters at the poles with bottoms permanently in shadow because its obliquity is only 0.034° so the Sun never rises to a high altitude in the sky above them. Here, it is always cold, and any water that makes it to Mercury could survive as ice almost indefinitely. First hinted at by 1991 radar analysis, space-probe imaging of Mercury shows that this is really happening at the planet's North Pole and by extension, its South Pole (Figure 5.9): Simple craters appear partially filled by new ice, either from comet impacts or escaped from the interior of the planet. ☞ Chapter 13. ☞

Describing the accessible surface of a planet is straightforward, but the bulk of a planet is hidden from view. However, in its last year of operation, MESSENGER orbited Mercury only 100 kilometers (65 miles) above the surface. Such proximity meant that unexpected changes in its course, due to variations in Mercury's gravitational field, could be used as a probe of the planet's interior, for instance, its density distribution.

Mercury is a dense planet with 5.4 gm/cm^3. Only the Earth is denser than Mercury, and this is simply because the Earth's greater gravity does a better job of compacting our planet. Thus,

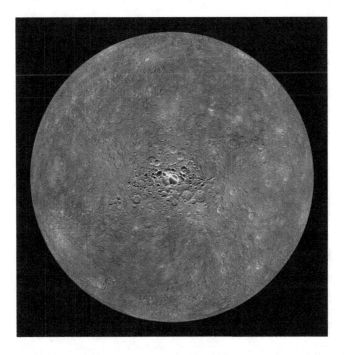

FIGURE 5.9   South pole of Mercury as imaged by MESSENGER.

Mercury must contain proportionately more of the higher density materials of which the inner Solar System is made and less of the lower.

On the surface of Mercury, we see rock. Underneath, the interior of the smallest planet consists of a lot of iron and nickel with a bit of sulfur. Three quarters of its radius consists of a metallic core. Mercury's remaining mantle and crust are only 400 kilometers (250 miles) thick. Picture a ball bearing frosted in rock. Well, not quite a ball bearing; Mercury's iron core contains within it an inner layer of liquid metal and possibly another inner-most solid sphere of the same. This is Mercury.

Why does Mercury possess such a large core? One thought is that it suffered an impact with another planet-sized object early in its history. The collision would have blasted off much of Mercury's mantle, leaving behind the iron core intact. Giant impacts appear to have been common while the planets were forming; such events also may have affected Venus, Uranus, and Neptune ☞ Chapters 6 and 11 ☞ Alternatively, an early hot Sun could have vaporized rocky materials in the mercurian crust. Or lighter materials suffered more drag in the inner proto-solar nebula and were lost from the material that accreted into Mercury. Each of these hypotheses predicts different surface conditions that will be tested by the BepiColumbo mission (Figure 5.10).

The internal temperature and pressure of Mercury is just enough to keep the interior liquid and to produce a weak magnetic field, which was detected by Mariner 10 and MESSENGER. This global field, which aligns with its rotational poles, is only about 1% the strength of the Earth's. Although weak, Mercury's field still holds off the strong solar wind.

Within Mercury, the liquid, metallic layer may provide the conducting material, but the planet's slow rotation seems insufficient to generate the electric current necessary to produce the measured magnetic field. A recent theory suggests that drops of liquid iron are solidifying and falling deeper into the planet's core, thereby providing the energy for the required currents. Imagine a 'snow' shower of iron flakes! While a similar process powers the Earth's magnetic field ☞ Chapter 7 ☞, the difference seems to be where the solidification occurs. In the Earth, solidification occurs at the boundary between the core and mantle while, inside Mercury, it happens at the boundary between the inner (solid) and outer (liquid) core.

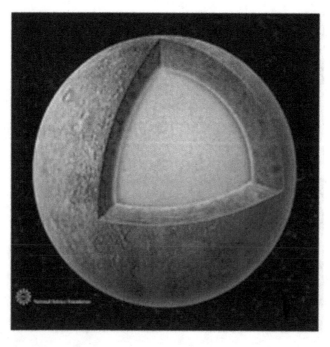

FIGURE 5.10   Mercury in cross section.

Even with its (relatively) large, dense metallic core, Mercury is still the least massive planet. It contains less than 6% of the mass of the Earth. Imagine standing on Mercury, encumbered by a space suit and backpack. However, because the surface gravity on the planet is only 38% of the Earth's, you can easily carry your increased mass in stride.

The European Space Agency [ESA] and Japan Aerospace Exploration Agency [JAXA] jointly launched BepiColumbo in 2018, which undoubtedly will tell us much more about Mercury when it arrives there in 2025. It is named after Italian celestial-mechanician Giuseppe 'Bepi' Colombo ⟨1920–1984⟩ who envisioned the gravity assist maneuvers used by Mariner 10 to flyby Mercury multiple times. First, though, the new spacecraft must slip into orbit around Mercury without plummeting uncontrollably into the immensely strong gravitational pull of the nearby Sun! BepiColumbo flew past Venus in 2020.

Although Mercury and the Moon are not as similar as they initially appear, neither are they as different as their separate 'planet' and 'satellite' classifications make them seem. As we continue to compare and contrast results from a variety of technologies, our understanding of these desolate, airless worlds, so different from our own experiences, increases.

NOTES

1 Mercury, the messenger, is still the symbol of the Florists Transworld Delivery [FTD] service.
2 Tragically, Arecibo's main radio telescope, spanning 305 meters (1,000 feet) in diameter, collapsed on 1 December 2020 due to a mechanical failure, ending nearly six decades of scientific discovery. It is currently decommissioned with no plans for repair.
3 He received the 1921 Nobel Prize in Physics for his contributions to theoretical physics, especially the photoelectric effect.
4 The IAU names mercurian scarps after famous ships, such as the USS *Enterprise*, which conducted surveys in the 19th century.
5 After RV *Belgica*, the first ship to spend a winter in the Antarctic.
6 The IAU usually names mercurian craters for deceased musicians, artists, and authors, such as Russian composer Sergei Rachmaninoff ⟨1873–1943⟩.
7 Water boils at 100°C (212°F).
8 Water freezes at 0°C (32°F).

6 Hothouse Venus

Did your parents teach you to recite, "Star light,/Star bright,/The first star I have seen to-night?" Would you have felt differently if you had known you were probably gazing at a planet: Venus, the Evening Star, the Morning Star, the Day Star.[1] At its brightest, Venus outshines all celestial bodies other than the Sun and Moon. It first appears in the east just before sunrise as the Morning Star. Over about the next 263 days, it moves gradually towards its maximum western elongation, or separation, from the Sun and back towards the Sun. Then, it disappears for about 50 days, only to reappear in the west as the Evening Star. Over the next 263 days, it moves towards its maximum eastern elongation from the Sun and back towards the Sun again. This time, it will disappear for only about 8 days before starting a new cycle as the Morning Star. Among others, the Greek mathematician Pythagoras ⟨circa 570 BCE⟩ appears to have worked out that this Morning Star and this Evening Star were a single body (Figure 6.1).

As early as 1702 BCE, the Babylonians were recording their observations of Venus on cuneiform tablets and predicting its future appearances. They associated the planet with their Queen of Heaven, Ishtar, a goddess both of love and beauty and of war and conflict. Through shared associations, the 'Star of Ishtar' became the 'Star of Aphrodite' (Greek) and, then, Venus (Roman).

The planet Venus was also significant to Mesoamerican cultures. The Mayans carefully tracked the motions of Venus, which they associated with their war god, Kukulcan. Several pages of the Dresden codex, one of three surviving Mayan books, are devoted to the venusian[2] 584-day cycle. They also aligned some of their buildings, such as El Caracol at Chichén Itzá in Mexico, towards the extreme horizon points of this cycle.

After millennia of naked-eye observations, Galileo turned his new telescope on Venus in 1610. Viewed through this instrument, Venus swelled from a mere point to a small disk. However, over time, Galileo saw Venus wax and wane through a full cycle of phases like the lunar phases: crescent, quarter, gibbous, and full. His observations lent support to the new but controversial Copernican model in which both the Earth and Venus orbit the Sun. The traditional Ptolemaic system with Venus orbiting the Earth could only produce crescent phases (Figure 6.2).

The Copernican model placed Venus closer to the Sun than the Earth. The 584-day cycle that fascinated the Mayans is the time required for Venus and Earth to move from a specific alignment, say Venus at its first appearance as the morning star, through their separate orbits around the Sun until their next such alignment. The time between such configurations of Venus, the Earth, and the Sun is the synodic year of Venus. If instead the position and motion of Venus is tracked with respect to the background stars, its orbital period of a mere 225 days emerges. Using Kepler's Laws, its distance from the Sun can be calculated to be 0.72 au, but how far it that really?

One disappearance of Venus occurs because it moves between the Earth and the Sun. However, if the conditions are just right, you can detect Venus as a black dot transiting the face of the Sun.

Transits of Venus are rare; after a pair of events 8 years apart, observers must wait more than 100 years (alternately 106 or 122 years) for another opportunity. Reviewing tables of planetary positions published earlier by Kepler, Englishman Jeremiah Horrocks ⟨1618–1641⟩ predicted the second Venus transit of the 17th century would occur on 4 December 1639. From his observations of that transit and measurement of the disk of Venus, he calculated the distance between the Sun and the Earth, or 1 au, to be about 15,000 times the radius of the Earth. He was a little low.

In hopes of setting an accurate scale for the Solar System, the international scientific community organized expeditions to view Venus cross the Sun in 1761 and 1769 and, again, in 1874 and 1882. Although each expedition contributed to a better estimate of this critical distance, transit

FIGURE 6.1   Venus in the Western sky, 24 November 2019. Venus is the brightest object, while the planet Jupiter shines above it.

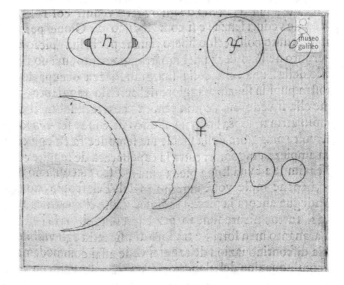

FIGURE 6.2   Page from Galilei's *Il Saggiatore (The Assayer)* showing Jupiter, Saturn, and the phases of Venus.

observations failed to achieve the level of accuracy and precision predicted by English astronomer Edmond Halley ⟨1656–1742⟩.[3] Observers reported a distortion of Venus as it touched the inner edge of the Sun's disk; several factors appear to contribute to the so-called black drop effect: diffraction, telescope optics, the Earth's atmosphere, and solar limb darkening.

By the 21st-century transits on 8 June 2004 and 5/6[4] June 2012, astronomers had accurately measured the Earth-Sun distance using other techniques; giving an orbital radius of Venus as 0.72 au or 110 million kilometers (67 million miles). Therefore, astronomers used these events as templates for future observations of extrasolar planets transiting their host stars. ☞ Chapter 16 ☞ New questions will be asked by a future generation of astronomers when Venus next transits the Sun in 2117 and 2125 (Figure 6.3).

With an orbital radius only 0.28 au less than that of the Earth, Venus comes closer to us than any other major Solar System body except for the Moon. Its nearness along with its very high albedo makes Venus an exceptionally bright object in our sky. Therefore, you would suppose astronomers would know a lot about it.

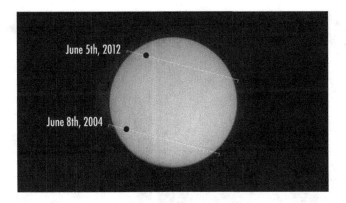

FIGURE 6.3 Two most recent transits of Venus.

When Galileo observed the Moon, he saw craters and other surface features. Tracking the motions of such surface features is one means of determining whether a body rotates and its rate. This method works well for Mars and most other planets. However, in a small telescope, Venus appears as a blank disk, or portion of one.

While observing the 1761 transit of Venus, Russian astronomer Mikhail Lomonosov ⟨1711–1765⟩ noticed the sunlight bending around Venus, as though its direction changed slightly when encountering an atmosphere. Almost two decades later, German Johann Schröter ⟨1745–1816⟩ realized that the points of crescent Venus extending farther than a half-circle could be explained by a scattering of sunlight. Therefore, unlike Mercury and the Moon, Venus has an atmosphere and a very thick one—90 times the pressure on the ground than we experience on the Earth (Figure 6.4).

Moreover, in visible light,[5] Venus is completely enshrouded in uniform, whitish clouds, which contribute to its high albedo. The human eye cannot detect distinct features, even with a good telescope. On a pool table, you can determine whether the balls are spinning fast or slow by watching the numbers appear and disappear. But the rotation of the white cue ball is another story; it might be spinning fast or slow—there is no way to say just by watching it. Therefore, the determination of the rotation period of Venus, a very basic parameter, had to wait another 171 years until radar provided a means of penetrating its clouds.

Venus sometimes is called 'Earth's twin.' With a diameter of 6,100 kilometers (3,800 miles), Venus is almost the same size, mass, and density as the Earth. However, the Earth is 5% larger,

FIGURE 6.4 Crescent Venus.

FIGURE 6.5   Earth (left) and Venus (right) to scale.

18% more massive, and, consequently, 5% denser. In addition, they both formed in the inner region of the Solar System. These similarities suggest their interiors should be comparable, too. However, no satellite accompanies Venus such as the Earth's Moon (Figure 6.5).

Starting in the 19th century, science-fiction authors freely speculated about what Earth's warmer twin might be hiding beneath its thick clouds. Adventure novelist Edgar Rice Burroughs ⟨1875–1950⟩ wrote of Amtor with its princesses needing rescue among tropical islands protected from the fierce Sun by the clouds while British author and theologian C. S. Lewis ⟨1898–1963⟩ describes Perelandra as a watery second Eden. Other theories included rain forests and dinosaurs![6] (Figure 6.6)

As delightful as these fancies are, improving radar technology eventually penetrated the cloud layers to reveal the surface of Venus. Just as an airline pilot can use radar to 'see' a runway in fog, astronomers use radio waves to penetrate the venusian clouds and interpret the signals that bounce back from the underlying surface. If a signal takes longer than average to return, it has encountered a valley. If it takes less time than average, it has found a mountain. The intensity of the returning signal also provides information about the nature of the venusian surface: Rough ground scatters the radio waves so less signal is returned while smooth ground reflects more of the incoming signal (Figure 6.7).

The rotation of Venus also causes radar signals to blueshift when they reflect off the limb that is approaching the Earth and to redshift when they reflect off the receding limb. These movement-induced changes in wavelength are called the Doppler Effect after the Austrian physicist who enunciated it, Christian Doppler ⟨1803–1853⟩. The Doppler Effect is more noticeable in sound waves, for instance the changing pitch of a car horn as a vehicle passes you.

Having waited all those years for radar to reveal the rotation of Venus, the result was surprising. Venus rotates once every 243 Earth days. This is extremely slow, especially when compared to its revolution period of 225 days. The 'day' on Venus is longer than the 'year!'

Every time Venus passes the Earth, known as inferior conjunction when they are closest, Venus shows the same face. The reason for this unusual 1.6-year occurrence is thought to be Earth's gravitational influence on Venus. In addition, the orbits of Earth and Venus arrive at *almost* the same configuration after 8 Earth orbits and 13 Venus orbits (an 8:13 spin-orbit resonance). Several

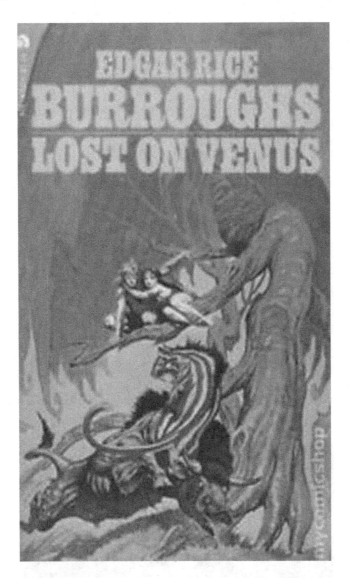

FIGURE 6.6 Venus in science fiction.

hypotheses have been put forth to explain this behavior but none are as yet definitive. Even more surprising, its direction of rotation seems backward to us—it is west-to-east. As seen from Venus, the Sun rises in the west and sets in the east.

Radar, however, is not the only tool revealing what lies beneath the venusian cloud cover. As they entered the space age, the USSR and US began sending spacecraft to explore this alien environment. The USSR launched the Venera series of 16 missions between 1961 and 1983, which provided detailed information about the planet, including the only landers so far. The US sent the Mariner series of ten missions between 1962 and 1973, which included three successful Venus flybys (Figure 6.8).

With all these similarities to the Earth plus its odd-sounding motions, what is it really like on Venus? It is not as cozy as the term 'Earth's twin' suggests.

One of the early indicators that Venus might be a dicey place to visit came when the first robot probes parachuted towards the surface—and then promptly *melted*. (The lead in their circuit-board solder did, anyway.) On 18 October 1967, Venera 4 parachuted into unseen Venus and became the

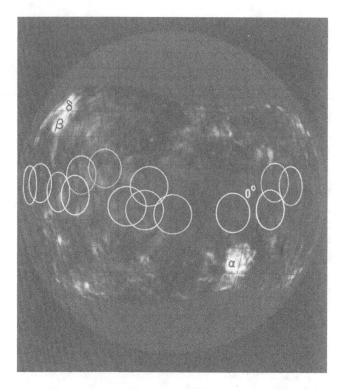

FIGURE 6.7   One of the earliest radar images of Venus made at the Arecibo radio observatory.

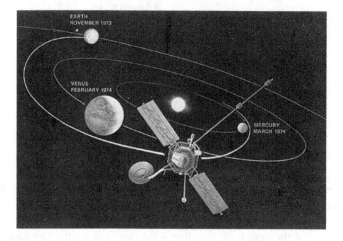

FIGURE 6.8   Trajectory of the Mariner 10 spacecraft taking it past Venus and Mercury in 1973.

first probe to experience the atmosphere of another planet; it succumbed to high temperature and atmospheric pressure after transmitting for only 93 minutes. These extreme conditions were consistent with those detected during the Mariner 2 flyby in 1962. Five years later, Venera 8 lasted 63 minutes on the inhospitable surface itself.

In 1975, Venera 9 detected acids in the venusian atmosphere, including hydrogen chloride and hydrogen fluoride; on Earth, these two chemicals are used in industrial processes, but they can produce severe chemical burns. Three years later, Venera 11 also identified sulfur and chlorine in the clouds. Each mission revealed a little more of the cloaking atmosphere or conditions beneath the clouds. The temperature on Venus averages 460°C (860°F), much hotter than a kitchen oven. Why is Venus so hot? Obviously, it is closer to the Sun than we are, but not as close as Mercury, and should that venusian atmosphere not have some moderating influence?

The venusian atmosphere is the problem. Unlike the Earth's air made of nitrogen and oxygen, the atmosphere of Venus is made up of almost entirely carbon dioxide, a colorless, odorless gas. We exhale carbon dioxide with every breath and drink it in every soda. However, the thick carbon dioxide atmosphere is nearly opaque to radiated heat making it unlike the atmospheres of the other planets. It is quite capable of setting up a runaway greenhouse effect (Figure 6.9).

Like a greenhouse, the venusian atmosphere lets in most of the Sun's visible light during the day, where it is absorbed by the planet's surface. When at night this energy is reradiated from the warm ground as infrared light,[7] it is trapped by the venusian atmosphere. The longer wavelength radiation cannot escape. The temperature rises, higher and higher. More carbon dioxide is baked out of the soil, contributing to the problem. A chain reaction of sorts occurs. The situation eventually reaches an equilibrium with the heat energy filtering out to space matching that of the incoming radiation. But, by then, the damage is done.

If a stifling venusian atmosphere is not enough, a fraction of it also contains sulfuric acid. On the Earth, we use this corrosive acid in automobile batteries, but, on Venus, it rains. This extreme acid rain evaporates before reaching the surface, which is bone dry. Although hurricane-force winds howl across the cloud-tops, hardly any breeze wafts across the surface. Constant lightning discharges may flare from the clouds. This vista is more evocative of a medieval description of Hell than an Earthly twin.

Unlike people, spacefaring robots can observe at wavelengths beyond those of visible light. In ultraviolet light,[8] for instance, moving sulfur-dioxide cloud features can be distinguished. The venusian atmosphere has no water clouds because sunlight breaks water molecules into hydrogen and oxygen atoms, which escape into space (Figure 6.10).

In 1985, the USSR missions Vega 1 and 2 each dropped a balloon and lander into the venusian atmosphere, which confirmed that, while the winds are absent at the surface, they are blowing at a fantastic rate in the upper atmosphere: 60 times faster than the planet rotates. The reason for this 'super-rotation' is unknown.

In the long daytime of Venus, its upper atmosphere absorbs a tremendous amount of heat, just as the surface does. That heat must go somewhere. When it rolls into night on Venus, what happens? In 2006, the ESA's Venus Express mission showed that nighttime cloud features use

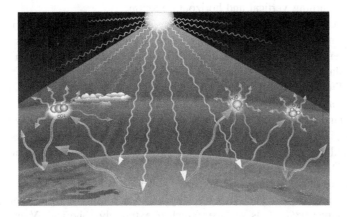

FIGURE 6.9   Greenhouse effect. Solar energy penetrates the planet's atmosphere, but reradiated heat is trapped.

FIGURE 6.10 Venus in ultraviolet light by Pioneer Venus. This false-color image shows the planet's sulfur dioxide clouds.

some of this excess energy daily to turn the hemisphere into an atmospheric mélange of changing cloud shapes and churning vertical and horizontal motion.

The Japanese space probe, Akatsuki ('Dawn'), has been orbiting Venus since its arrival in 2010. It carries instrumentation to study atmospheric composition, dynamics, and cloud structure. In 2017, mission scientists announced the discovery of an equatorial 'jet stream' in the upper atmosphere. Also launched with Akatsuki, JAXA's Interplanetary Kite-craft Accelerated by Radiation Of the Sun [IKAROS][9] successfully flew by Venus in 2010 using solar sail technology, the first such demonstration of propulsion by sunlight in interplanetary space.

Some Venera landers survived long enough to send back a few pictures of Venus. The view is of a stone-strewn plain to the horizon. The stones look comparatively smooth because of chemical erosion from the relentless atmosphere. Indeed, the rocks of Venus resemble many of those on the Earth (Figure 6.11).

To examine more of the surface of Venus, we still must rely on radar, now incorporated into space missions visiting Venus. The NASA's Magellan spacecraft went into orbit about Venus in the 1990s and mapped 98% of its surface, which is extremely spherical. Venus turns out to be somewhat flatter than the Earth: While six mountainous regions cover 1/3 of Venus, from the

FIGURE 6.11   Surface of Venus from Venera. The fish-eye lens shows the foot pad of the lander and the distant horizon.

deepest valley to the tallest mountain (Maxwell Montes), the relief on the planet is only about 14 kilometers (8.7 miles). The two continents on Venus are not lands rising above a sea level but rather above a mean level of elevation. Whether they resulted from plate tectonics or their borders ever were shores of a venusian ocean is a question that cannot yet be answered definitively. Regardless, if it is true, the ocean boiled away long ago (Figure 6.12).

A planetary surface that is completely cratered is probably a geologically dead world. One on which erupting volcanoes dominate is a geologically active one. Venus might be some sort of intermediate case.

Venus has far fewer craters than Mercury or the Moon. What craters there are tend to be larger ones. Many bodies that might form craters on Venus never make it through the planet's dense envelope. However, throughout the history of Venus, some large meteoroids reached the surface and produced craters. Because most of the detectable ones appear new, the surface may have undergone a bout of crater-erasing lava flows (Figure 6.13).

In addition, Venus has round (conical, actually) features that appear to be volcanoes—tens of thousands of them—some with what looks like material that has flowed down their sides. Are the venusian volcanoes extinct or have they erupted recently?

ESA's Venus Express orbited Venus from 2006 through 2014.[10] During its lifetime, it picked out hot spots coincident with putative venusian volcanoes. Moreover, its instruments measured trace changes in the atmospheric chemistry, of a kind that might be expected after the release of gases from a volcano—within the past year! No 'smoking gun' proves current volcanism on Venus, but we may not be far from discovering it (Figure 6.14).

Maxell Montes is the highest mountain on Venus at 11 kilometers (6.8 miles) above the mean surface. It has a large crater, named Cleopatra, with a diameter of 110 kilometers (65 miles), on its *slope*. Whether this particular peak is a volcano is still unclear, but, if it is, the coincidence of a volcano and impact crater ending up in the same place speaks to the plentitude of both features on Venus.

Right now, planetary scientists find no conclusive evidence of global geologic activity on Venus such as on our home planet. Its crust does not appear to have breaks that might allow liquid material from its mantle to spread across and recoat its exterior. ☞ Chapter 7 ☞ Perhaps the

FIGURE 6.12   Venus in radar by Magellan.

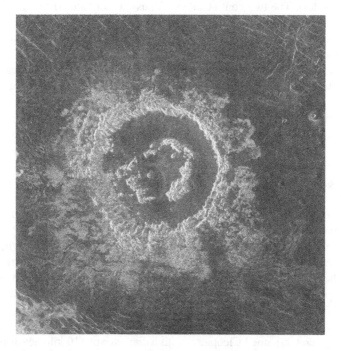

FIGURE 6.13   Magellan radar image of a large impact crater on Venus.

venusian crust is too hot and malleable. However, the low crater density on Venus points to some sort of global geologic activity in its past, perhaps as recently as 250 million years ago. Whether this was a catastrophic volcanic resurfacing of Venus or a more gradual steady process is a matter of controversy among planetary geologists.

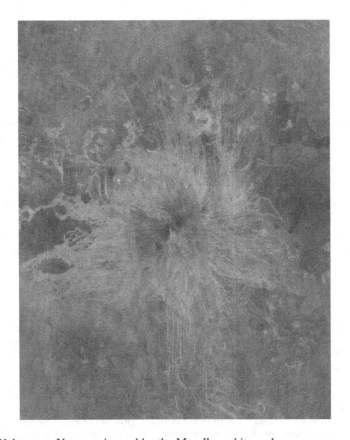

FIGURE 6.14   Volcano on Venus as imaged by the Magellan orbiter radar.

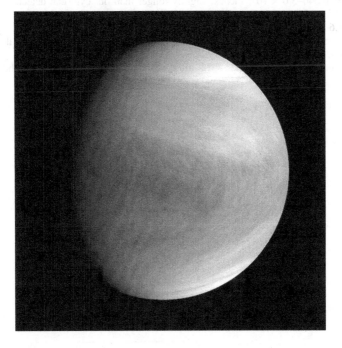

FIGURE 6.15   False-color image of Venus by Akatsuki.

Other local geological formations, tantalizingly suggestive of geologic activity and found uniquely on Venus, include coronae and tesserae. Coronae are blistering of the venusian crust caused by rising material beneath it. Tesserae are similar but crisscrossed by ridges and valleys. If substantiated, the presence of current geologic activity on Venus may ultimately prove to be that planet's greatest claim to kinship with Earth.

With a density of 5.2 g/cm^3, we expect an iron metal core lies beneath the crust and mantle of Venus. However, with its extremely long rotational period, Venus has no intrinsic magnetic field. The solar wind impinges directly on the atmosphere, creating a layer of ions that induces a unique magnetic environment, which was also studied by Venus Express.

Sending space probes to distant worlds is expensive. In the case of Venus, these large expenditures not only feed our innate urge to explore, but also provide a cautionary tale: We are inadvertently changing our climate on the Earth by making the atmosphere more Venus-like. In June 2021, NASA announced two missions, DAVINCI+ and VERITAS, to study Venus's hothouse climate and other outstanding issues. Scientists studying terrestrial climate change have a worst-case scenario in Venus. By studying Venus, we can perhaps help avoid a similar fate for the Earth (Figure 6.15).

NOTES

1 The authors try to reserve our wishes for actual stars but are occasionally overcome by enthusiasm.
2 Cytherean is also used to describe something related to the planet Venus; the term comes from Cytherea, a poetic name for the Greek goddess Aphrodite, who supposedly first came ashore on the island of Cythera.
3 Halley is better known for the comet that bears his name. ☛ Chapter 13 ☚.
4 The transit began at 22:09:41 on 5 June and ended at 04:49:31 on 6 June. The times are given in Universal Time, Coordinated [UTC]. The time and day would vary depending on your local time zone.
5 Our eyes are tuned to see electromagnetic radiation, or light, with wavelengths between 400–750 nanometers (1.6–3.0×10^{-5} inches), which is the visible region of the electromagnetic spectrum.
6 In 2020, the tentative discovery of atmospheric phosphine, a potential marker of life, reinvigorated the debate about life on Venus. ☛ Chapter 17 ☚.
7 Infrared radiation is a form of electromagnetic radiation, or light, with wavelengths 1 millimeters– 750 nanometers (0.040–3.0×10^{-5} inches, slightly longer than our eyes can detect.
8 Ultraviolet light is a form of electromagnetic radiation, or light, with wavelengths 10–400 nanometers (3.9×10^{-7} – 1.6×10^{-5} inches), just shorter than our eyes can detect.
9 Which sounds like Icarus, who, in Greek mythology, died when he flew too close to the Sun on wings of wax and feathers.
10 Its science mission ended in December 2014, after which it probably burned up in the atmosphere as its orbit decayed.

7 Terra, Our Earth

The 'Blue Marble' photograph taken by Apollo 17 astronauts *en route* to the Moon transformed our image of the Earth. *The Book of Common Prayer* evokes "this fragile earth, our island home." However special the Earth may seem to us, and to the billions of other living creatures inhabiting it, our home is fundamentally a planet like the others; it is, as Joe Diffie sings, the "third rock from the Sun."

This book on the Solar System strives to be fair, giving it no more or less attention than its companions orbiting the Sun. Yet, we have so much more knowledge about the Earth simply because we happen to live here. Keeping to our planetary science context, we will consider some of the ways in which the Earth is similar to and different from the other terrestrial planets (Figure 7.1).

For one thing, scientists have a more complete model of the interior of the Earth than they do of the other planets, largely through seismology. A world-wide network of seismometers has been detecting tremors moving through the Earth since Cecchi started recording earthquakes in 1875. Primary [P] waves push buildings one way and, then, pull them back along the surface of the Earth. Secondary [S] waves shake the terrestrial surface either back and forth or up and down.

Although the waves observed are similar to those captured by the Apollo seismographs ➤ Chapter 4 ➤, the more extensive and continuously terrestrial-monitoring system allows the development of a more detailed and better substantiated model of the Earth. This model consists of a central core, surrounded by a mantle and then an outer crust. As a P wave crosses from the mantle into the core, its speed and direction change. Then, its speed and direction change again as it moves from the core into the mantle. Meanwhile, S waves travel nearly through the center of the Earth only to stop entirely, indicating they have encountered a liquid. From this, geologists deduce that the mantle is plastic, but the outer core of the Earth must be fully liquid and the inner core solid. Swathed in the near-frictionless outer core, the inner core may rotate even faster than the planet as a whole.

The terrestrial density of 5.5 gm/cm^3 implies that the Earth consists of rock and metal. Seismic maps describe how those materials are distributed. If we could cut the Earth in half, the cross section would reveal a thin, outer crust of low-density rock covering a viscous mantle of high-density rock, which itself surrounds a metallic core of iron and nickel (Figure 7.2).

The Earth's metallic core is growing because iron flakes within the mantle continue to condense and rain down upon it. However, most of the Earth's mass remains in the mantle.

Why are the terrestrial worlds layered (or differentiated), separated into distinct horizontal regions consisting of distinct materials? When these planets formed, they were hot and molten. The lower-density materials floated to the top (the outside of a sphere) while the higher-density materials sank to the bottom (the center of a sphere). Depending upon how you brew your coffee, you may see a similar process whereby the heavy grinds sink to the bottom of your cup while a thin sheen of oil may form on the surface.

As miners move downward into the Earth along a mineshaft, they notice their surroundings get hotter. Extrapolating, the planet's core must be very hot, part of it a shell of liquid. The mantle is less fluid, but still consists of semi-molten, churning rock.

Volcanoes provide dramatic and, sometimes, disastrous evidence of conditions in the mantle. Liquid rock emerging from a volcano must come from a place hot enough to melt rock. The absence of many volcanoes on Mercury and the Moon points toward their having cooler interiors and less geologic activity. Understanding the volcanoes of Venus would indicate the level of geologic activity there and the internal heat driving it.

FIGURE 7.1   Earth.

FIGURE 7.2   Earth in cross-section.

Meanwhile, over a thousand volcanoes on the Earth simmer (Figure 7.3). New ones continue to be discovered, such as the hitherto unknown volcano that surfaced above the Indian Ocean in 2018.

The terrestrial crust is uniquely broken into chunks, or plates, that may move under the influence of convection in the mantle below them. Convection distributes heat throughout a pan of water boiling on a stove; water heated by the burner beneath rises while colder water from the

FIGURE 7.3   Hawai'i's Kilauea volcano erupts in 2017.

surface sinks. Similarly, the bottom of the mantle is itself warmed by the even hotter core beneath it. Heated lower-density rock from the bottom of the mantle slowly rises towards the top, where it cools before sinking. The same phenomenon makes watching lava lamps so fascinating!

Plate tectonics is the slow shifting and re-arranging of the terrestrial crust. Where two plates meet, one may be forced down into the mantle. Such subduction zones are usually marked by ocean trenches and volcanoes. Along the Cascadia Subduction Zone, a dense ocean plate is sinking below the Northwestern USA and Southern British Colombia, producing the volcanoes of the Cascade Mountains. Mount Saint Helens erupted explosively in 1980 and nearly continuously from 2004 to 2008. Colliding plates also may fold and bulge giving birth to mountain ranges such as the Himalayas.

In addition, the descending slab pulls its plate away from the plates along its other edges. In some places, separating plates produce chasms, such as the complex system known as the African rift valley. Elsewhere, molten rock from the mantle spreads upward creating ridges pushing the plates away. The Mid-Atlantic Ridge is spreading, allowing mantle material to emerge, and creating 1–10 centimeters (0.4–4 inches) of new seabed each year. Swelling up to 3 kilometers (1.9 miles) above the surrounding ocean bottom, the crest also contains an underwater rift valley. The dual processes of subduction and rifting are continuously recycling the Earth's crustal material.

Why plate tectonics? We do not know; one possibility is that the pattern of plate boundaries is left over from cracks formed when the Earth's crust solidified. ➤ Chapter 1 ➤ (Figure 7.4)

A volcano can form wherever a weak spot in the mantle allows liquid rock to force its way up through the crust. Eventually, though, plate tectonics moves this piece of crust away from the hot spot, causing the volcano to become dormant and cease growing bigger. However, the hot spot is now beneath another piece of crust, where it produces another volcano until this section of crust moves aside. Then, another volcano forms and another and… A chain of volcanoes is created, running, as if on a conveyor belt, parallel to the direction the crustal plate is traveling.

An example: The Hawai'ian Islands are a chain of volcanoes strung out as the Pacific Ocean plate moves northwest. They are the most active in the world. The ones farthest to the northwest are so aged and eroded that they are mostly below the ocean once again. Of the ones eastward, Kaua'i is old and rugged while Maui still looks like a volcano but appears to be extinct. However, on the Big Island of Hawai'i, now seated above the hot spot, volcanoes are very active. Kilauea erupted nearly continuously from 1983 to 2018; its flowing lava adding nearly 2.3 square kilometers (0.89 square miles) of land to the island while destroying more than 200 buildings and 14 kilometers (9 miles) of roads. Eventually, the Big Island will move away from the hot spot and become inactive while a new volcano will develop. This process has started already. Lō'ihi, a fresh

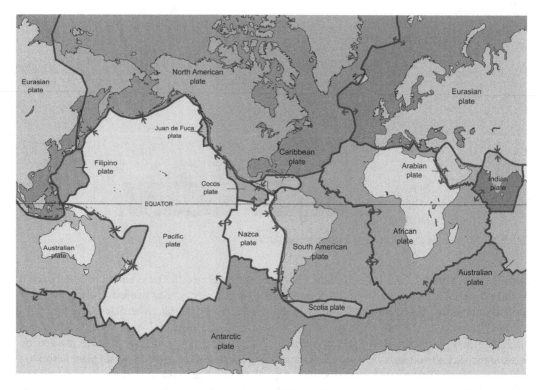

FIGURE 7.4   Map showing plate boundaries in the Earth's crust.

volcano east of Hawai'i, is not above ocean level yet, but once it is, people will probably flock to it as a pristine, tropical paradise! (Figure 7.5)

Another consequence of moving plates is earthquakes. When two plates slide along one another, they may stick temporarily; the sudden release of energy when they become unstuck results in earthquakes, which send P and S waves throughout the Earth. Lying between the Pacific Ocean plate and the North American plate, the San Andreas Fault produced the great 1906 earthquake that destroyed San Francisco as well as many lesser events. In 2010, another fault caused an earthquake that killed hundreds of thousands in Haiti (Figure 7.6).

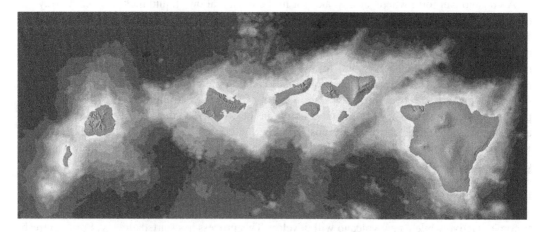

FIGURE 7.5   Hawai'ian volcanic chain. Lö'ihi eventually will appear in the lower left corner of the map.

FIGURE 7.6   Haitian earthquake damage.

An earthquake in the sea may generate a powerful wave called a tsunami. When it came ashore, the 2004 tsunami in the Indian Ocean killed hundreds of thousands more. With its volcanoes, earthquakes, and tsunamis, the Earth is a dangerous place, but only a geologically active world would likely become the home of life.

As we write, geologic activity and erosion resurface the Earth completely every 600 million years or so. On Earth, erosion is primarily caused by water flowing on and through the crust and by weather generated in the atmosphere that blankets the planet.

The Earth's atmosphere consists of four layers, the troposphere; above it, the stratosphere; above it, the mesosphere; and above it, the thermosphere (Figure 7.7). Because the same processes are at work on Venus, it has these same atmospheric layers, though they occur at different altitudes and, of course, are different temperatures as a function of height.

The troposphere, from the Earth's surface to about 9 kilometers (6 miles) in altitude, is home to weather. The Sun heats the Earth's surface, which in turn heats the bottom of the troposphere. Warm parcels of air rise; they are less dense than those of cool air. As these parcels travel farther from their heat source, they cool. They become denser and sink. This up-down motion is a primary driver of weather and is another example of convection. It works horizontally, too. Wherever air pressure is higher than its surroundings, air will travel from 'high' to 'low.' The result is wind.

Globally, there is a steady flow of air from the cool poles to the warm equator. However, the Earth is a rotating sphere, so the surface beneath this flow moves at different speeds causing it to veer east or west, breaking into patterns known as the jet streams.

Water gets into the weather act, as well. Liquid water in oceans, lakes, and puddles is heated by the Sun until it evaporates upward into the atmosphere. This now water vapor cools and condenses into clouds. Eventually it falls back to the Earth as rain, snow, or hail, which begins the cycle again (Figure 7.8).

Carbon dioxide high in the troposphere is partly responsible for a mild greenhouse effect. Unlike on Venus, the result is beneficial for life: Without greenhouse warming, the average temperature on the Earth would be below 0°C (32°F). The natural warming hikes the Earth's temperature by about 15°C (27°F) to a comfortable 15°C (59°F). However, burning fossil fuels releases additional greenhouse gases, primarily carbon dioxide but also methane. As tropospheric carbon dioxide levels swell, the average global temperature also increases.

Beyond the troposphere, the stratosphere extends to an altitude of about 50 kilometers (30 miles). Here sits the ozone layer, consisting of molecules built from three oxygen atoms each. This repository absorbs harmful ultraviolet light from the Sun and protects life. In 1982, satellite images revealed especially thin regions in the ozone layer above the poles. Atmospheric scientists

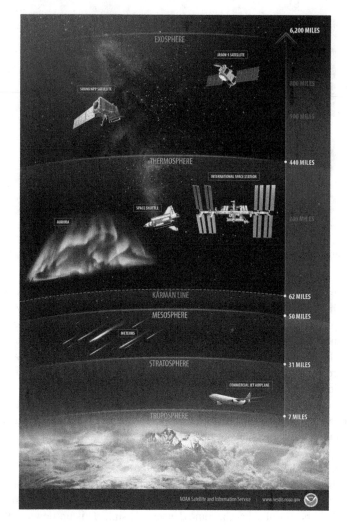

FIGURE 7.7 Diagram showing layers of the Earth's atmosphere.

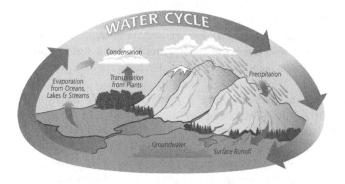

FIGURE 7.8 Diagram showing the Earth's water cycle.

worried that the holes might widen and eventually decompose the ozone layer. However, in 1974, chemists, Mexican-American Mario Molina and American F. Sherwood Rowland[1] ⟨1927–2012⟩, had demonstrated that the chemicals used mostly in refrigerators and air conditioners, chlorofluorocarbons [CFCs] like Freon®, could damage ozone in the atmosphere. When these chemicals are released into the air, they slowly ascend to the stratosphere. Ice particles over the poles facilitate ozone's demise. With global cooperation in 1987, CFCs were phased out of production, and the ozone layer is slowly recovering.

How important is reversing the ozone depletion in the Earth's stratosphere? We will find in the next chapter that Mars has virtually no oxygen in its atmosphere, so solar ultraviolet light shines unimpeded on its sterile surface. And it is but one example of how studying another planet helps us to better understand our own.

Because solar energy is absorbed there, the temperature in the stratosphere actually increases with altitude. This temperature inversion prohibits the convection prevalent in the troposphere. The stratosphere has no weather, so reaching the stratosphere means a smooth trip for an airliner.

The atmosphere cools with altitude again in the mesosphere ending 90 kilometers (60 miles) up. However, the air there is extremely rarified.

Thankfully, dangerous radiation from space is absorbed in the still-thinner thermosphere, which warms a bit due to its efforts. Some earth-orbiting satellites do so *within* the thermosphere.

Although the liquid, outer core of the Earth lies 2,900 kilometers (1,800 miles) beneath us, it generates a magnetic field. This field helps protect the atmosphere, and ultimately the surface, from the solar wind. A strong magnetic dynamo is at work here. The metallic core of the Earth provides the conducting material, which like the mantle above it has convective currents moving through it. Those currents, along with the relatively rapid rotation of the planet, put internal electricity in motion.

The other terrestrial worlds lack such strong magnetic fields. The lunar core has a liquid, metallic outer layer but is a slow rotator; hence, no magnetic field exists now, but lunar rocks record the presence of one in the distant past. Although Mars rotates quickly enough and contains a liquid, metallic core, it no longer has an intrinsic magnetic field. Venus does not generate its own magnetic field; it may have a liquid interior but lacks the rapid rotation thought necessary. Mercury is the exception that proves the rule: Its planetary dynamo is just barely working. In the outer Solar System, Jupiter is a massive, fluid planet and rotates rapidly; not unexpectedly, it boasts a strong magnetic field, like the other giant planets.

The Earth's magnetic field lines converge at the two oppositely aligned magnetic poles. On a bar magnet, we label these 'north' and 'south.' If you ever play with two magnets, you will notice that two north poles repel one another while a north pole and south pole attract one another. Despite the names, the terrestrial magnetic poles do not correspond exactly to its geographic poles. By 2020, the magnetic axis differed from the rotational axis by 9° and did not pass through the center of the Earth. Recently, the magnetic poles have been wandering 50 kilometers per year (30 miles per year) in response to currents within the terrestrial core and electric currents in the Earth's atmosphere. Right now, the North Magnetic Pole is headed for Russia. Eventually, the magnetic poles will switch from North to South, in the process reducing the magnetic field to 10% its normal strength. The geological record reveals that reversals occur on average every million years or so. However, we cannot be certain when the next reversal will take place. Today, though, mariners can navigate by compass needle, correcting for the difference between true north and magnetic north (Figure 7.9).

Energetic particles trapped by the terrestrial magnetic field are harmful radiation. Zones around the Earth with dense concentrations of this radiation are named the Van Allen Radiation Belts, after American physicist James Van Allen ⟨1914–2006⟩ who discovered them using a radiation detector aboard the first US artificial satellite, Explorer 1, in 1958. They have practical significance. Astronauts traveling to the Moon, for instance, must not spend too much time in these Belts lest they suffer from biological damage by radiation exposure (Figure 7.10).

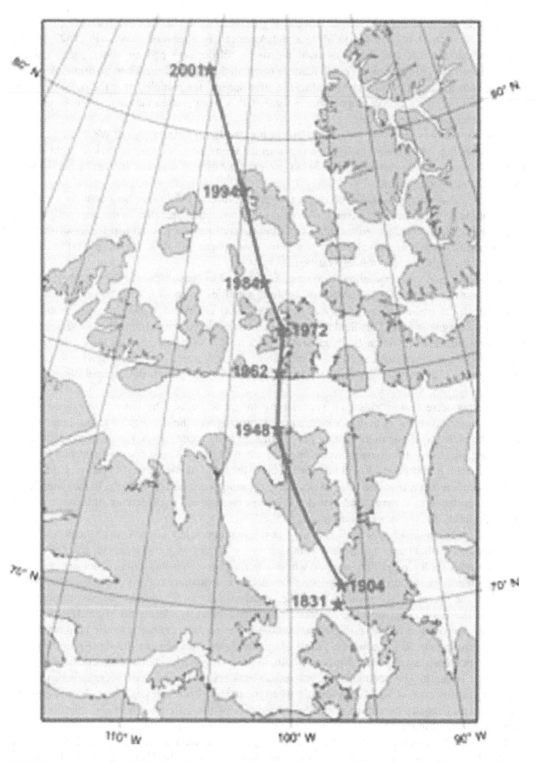

FIGURE 7.9 Map showing the Earth's wandering north magnetic pole.

FIGURE 7.10   Diagram showing the Van Allen Radiation Belts.

Most of the time, the terrestrial magnetic field protects us from this hazardous radiation by trapping it before it reaches the surface. Charged particles move up and down the field lines toward where these lines intersect the Earth: the magnetic poles. There, the only thing standing between harmful radiation and us is the atmosphere. As charged particles strike the air, the molecules absorb the particles' energy and reemit it as harmless light. Depending upon how close to the poles you live, you will see this light as a colorful glow at your northern or southern horizon. In the Northern Hemisphere, we call it the *aurora borealis* but, more commonly, the 'northern lights.' In the Southern Hemisphere, it is the *aurora australis* or 'southern lights' (Figure 7.11).

Beyond the dancing lights of the aurora, the Moon is locked in its own dance with the Earth. The phases of the Moon ➤ Chapter 4 ➤ describe how the lunar appearance changes as we watch it turn slowly on its axis while orbiting the Earth each month. Astronauts viewing the Earth from the Moon see the Earth constantly in their sky waxing and waning through its own phases, which are opposite to the lunar phases seen by their families at home.

The relationship between Earth and Moon is an intimate one that goes well beyond appearances. Consider three things: the Earth, an ocean on the side of the Earth closest to the Moon, and an ocean on the opposite side. Because they are liquid, the oceans flow. The near ocean is closer to the Moon than either the center of the Earth or the far ocean. Because gravity decreases with distance squared, a *stronger* force is tugging on this near ocean than is pulling on the Earth. This greater force causes the near ocean to be pulled away from the Earth. Meanwhile, the gravitational force that the Moon is exerting on the far ocean is even *weaker* that the force being exerted on the Earth. This time, the force differential pulls the Earth away from the far ocean. The result in each case is the same: The middles of both oceans seem to be pulled out and away from the Earth. At any given time, such tidal bulges exist on opposite sides of the Earth. The ocean is, in fact, slightly deeper in the bulge. Where does the extra water come from? At the same time, other places on the Earth[2] exist where the ocean and the center of the Earth are equidistant. There is no gravity differential here, but because water flows and all the oceans are connected, these parts of the oceans lose water to accommodate the tidal bulges. In these parts of the ocean, the water is slightly shallower than average, and a tidal trough is present (Figure 7.12).

Remember that the Earth is rotating and does so much faster than the Moon orbits the Earth. The tidal bulges remain on a line connecting the centers of the Earth and the Moon. Therefore, as the Earth turns, different parts of its surface experience a tidal bulge. From our point of view on the

FIGURE 7.11 *Aurora borealis* reflected in a glacial lake.

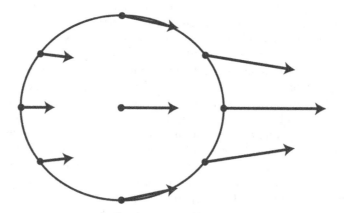

FIGURE 7.12 Diagram showing the differential pull of the Moon's gravity on the Earth.

Earth, the tidal bulges seem to be moving around Earth once every day or so. Say, now, that you are on an island or seashore that is rotating with the Earth into the tidal bulge. The water gets deeper and comes up higher onto the land; the tide comes in. As your tract of land rotates out of the tidal bulge and into the trough, the water level goes down; the tide goes out. You shortly encounter the second tidal bulge and the second trough. The same sequence happens. Then, the cycle begins again. Thus, in a little more than a day, every place experiences two high tides and two low tides. If you happen to be at the end of a narrow bay near where the water of the incoming tide piles up, the effect is magnified (Figure 7.13).

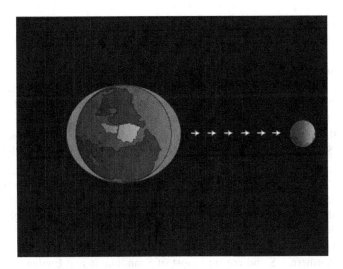

FIGURE 7.13   Diagram showing lunar tides.

The Sun causes tides, too. It does so in just the same manner as the attraction between the center of the Earth and the center of the Moon. The Sun is much more massive than the Moon, but it is also much farther away. Therefore, its gravitational force, and hence the gravitational differential it causes, is less than that of the Moon. Still, the net tide is the sum of the lunar and solar tides. When the Moon, Sun, and Earth are aligned during the New Moon or Full Moon, their effects combine to produce a high tide that is higher and a low tide that is lower than usual; this is a spring tide. When the Moon, Sun, and Earth are 90° from one another during First or Last Quarter Moon, their tides partially cancel each other to produce a high tide that is lower and a low tide that is higher than usual; this is a neap tide.

The Moon causes a tidal bulge in the solid Earth as well, albeit one of only 30 centimeters (10 inches), compared to as much as a meter (40 inches) of water in the deep ocean. The difference occurs because the tensile strength of the rock is so much greater than that of the water. Because the solid Earth does not flow as well as its oceans, this tidal bulge cannot slip around the Earth as well, either. As the Earth rotates, it carries the bulge with it. It takes time to stretch and bend the Earth. It takes time for the new bulge on the Earth-Moon line to form and the old bulge now rotated off the Earth-Moon line to dissipate. Because of this time lag, the tidal bulge of the solid Earth is always ahead of where it ought to be—that is, it is not directly below the Moon.

Consequently, the tidal bulge acts as a lever arm on the Moon. Gravitational attraction between the Moon and the bulge is no longer acting in the same direction as the attraction between the center of the Earth and the center of the Moon. Nevertheless, the two are drawn to each other. The Moon tends to be pulled forward toward the bulge, slightly accelerating the Moon into a higher orbit, and increasing the lunar separation from the Earth. At the same time, the Moon tugs on the bulge in the direction opposite the terrestrial rotation, applying a torque to the rotating Earth. The Earth slows down! This effect is miniscule, but over millions of years it adds up. The day is now noticeably longer than it was 100,000 years ago.

The seasons and climate are also changing. The Earth's geographic, or rotational poles, are tilted 23° from straight up and down with respect to its orbital plane, the ecliptic. The tilt is essentially fixed in space as the Earth orbits the Sun. This obliquity is the reason we experience spring, summer, autumn, and winter.

As the Earth moves around the Sun, its Northern Hemisphere leans toward the Sun. During this time, the sunlight strikes this region more directly and, consequently, heats the surface more effectively. Simultaneously, the Sun also appears above the horizon for a greater fraction of the day allowing it more time to heat the surface. On the longest day of the year known as the summer solstice, the Sun rises at northernmost point along the eastern horizon. At noon on that day, it will also reach its maximum height above the horizon. A few hours later, it will set at its northernmost point in the west. Summer has officially come to North America.

At the same time, South America is experiencing winter as it leans away from the Sun. During this interval, the sunlight striking the Southern Hemisphere is spread out and diffused more so that it heats the surface less effectively. Simultaneously, the Sun also appears above the horizon for a shorter fraction of the day allowing it less time to heat the surface. On the shortest day of the year, even the noonday Sun does not rise very high in the sky. Winter has officially come to South America.

As the year progresses, the seasons change: Northern Hemisphere summer giving way to winter weather, and Southern Hemisphere winter into summer weather. About halfway between the solstice extremes, the Earth experiences periods of roughly equal day and night, the equinoxes (Figures 7.14 and 7.15).

The weather patterns today are not as they have always been. More than once, glaciers extended as far south as the northern US and covered most of Scandinavia and England.[3] The evidence of past ice ages is written in land features and rocks, in the growth patterns of stalactites, in sediment accumulated on the ocean floor, and snowfall layered into Arctic ice sheets. In 1941, Serbian mathematician Milutin Milankovitch ⟨1879–1958⟩ identified three astronomical cycles affecting terrestrial climate: eccentricity, obliquity, and precession.

Over a 100,000-year cycle, the orbit of the Earth changes shape slightly; it ranges from more circular than it is now to a more distinctly elliptical shape. When the terrestrial orbit is more eccentric, the amount of sunlight, or insolation, received at different times of the year varies more than at present (Figure 7.16).

Over a 40,000-year cycle, the obliquity of the Earth changes slightly from 22° to 25°; it is currently near the middle of this range. The changes in axial tilt would be more extreme but the Earth has a rudder that helps steady its obliquity—the Moon is a stabilizing influence, which may

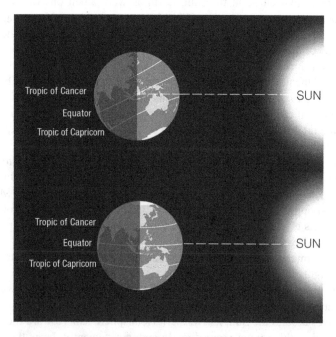

FIGURES 7.14 Diagram seasons in the Northern Hemisphere: Spring (top) and Autumn (bottom). [Seasons in the Southern Hemisphere: Autumn (top) and Spring (bottom).]

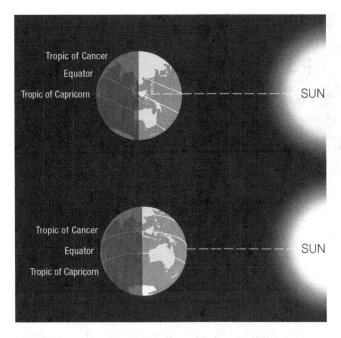

FIGURES 7.15   Diagram showing seasons in the Northern Hemisphere: Summer (top) and Winter (bottom). [Seasons in the Southern Hemisphere: Winter (top) and Summer (bottom).]

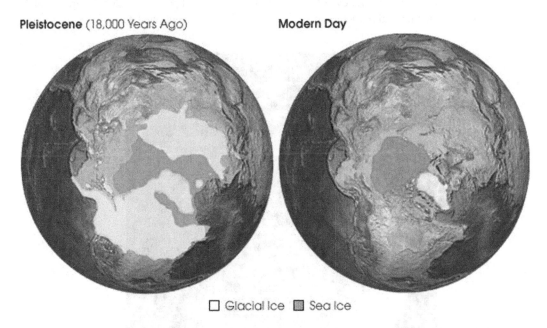

FIGURE 7.16   Map showing glacier coverage of the Earth, 18,000 years ago (left) and now (right).

keep the terrestrial axis from flipping around. With the obliquity of the Earth reduced, the distribution of sunlight across the planet is more even and seasonal variation is lessened (Figure 7.17).

Over a 26,000-year cycle, the direction in which the terrestrial rotation axis points also changes. Precession is the slow shift of this axis around the vertical, like a top that is set spinning slightly off kilter. Today, the North Pole points to Polaris, a star in the constellation Ursa Minor, the Little

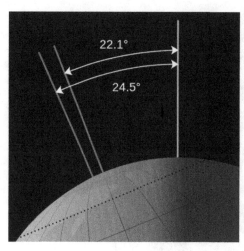

FIGURE 7.17 Diagram showing the changing obliquity of the Earth.

Bear. However, when the Egyptians built the pyramids, Thuban, in Draco, the Dragon, was closest to where the North Pole was pointing. This wobble changes the relationship between the winter solstice and the day on which the Earth is at perihelion. Currently, the winter solstice for the Northern Hemisphere occurs within a few weeks of perihelion, which produces milder winters. If the winter solstice occurred closer to the day on which the Earth is at aphelion, the winters would be longer and more extreme (Figure 7.18).

As a consequence of these natural cycles, the Earth has experienced periods of global warming and global cooling recorded throughout its geological past. However, as mentioned earlier, the amount of

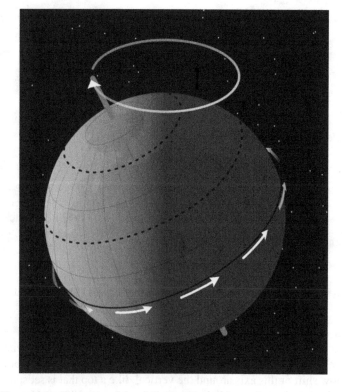

FIGURE 7.18 Diagram showing precession of the Earth's rotation axis.

carbon dioxide, a greenhouse gas, in the troposphere is increasing. Since the mid-20th century, climate scientists have observed unprecedented increases in both actual average temperatures and the rate of change of those temperatures. Overall, temperature records reveal an overall atmospheric temperature increase of about 0.7°C (1.3°F) since 1981, which is a rise of about 0.18°C (0.32°F) per decade over the same period.

Between 1990 and 2014, the Intergovernmental Panel on Climate Change [IPCC][4] issued five reports concluding that *human* activity in the industrial era is the primary cause of recent climate change.[5] Their findings are not disputed by any scientific body of national or international standing. In addition to rising atmospheric temperatures, climate change includes shifting precipitation patterns and rising sea levels with associated flooding as well, among other disastrous effects.

In addition to burning fossil fuels, agricultural emissions and deforestation are significant sources of greenhouse gases. Changing the chemical makeup of the troposphere is the greatest human impact on the Earth. As the temperature rises in response, the Earth experiences less snow cover, increasing water vapor, and melting permafrost, which exacerbate the climate problem.

Extreme heat, conditions that are hotter or more humid than is typical for a location, is one of the most fatal natural hazards. It poses grave risks to human health. Climatologists attribute trends in extreme heat to the influence of anthropogenic greenhouse gas emissions. As these gases continue to accumulate in the atmosphere, climate scientists project extreme heat to become more frequent and more severe. Imagine a weather forecast for tomorrow's highs in the 50s°C (120s°F) in the not-too-distant future. Now imagine a family 'sheltering in place' for months to escape the oppressive heat! Our lives will inalterably change unless we take action.

Climate change is one of the greatest challenges to our 21st-century civilization. Our successful mitigation of the threat posed by ozone depletion earlier is encouraging as we approach the problem of slowing or reversing climate change. Mitigation efforts include reducing greenhouse gas emissions from fossil fuels, switching to renewable energy technologies,[6] preserving and restoring forests, and geoengineering to positively affect the Earth's climate. Will we have the political and economic will in the short term to act on limiting climate change in the long term before our civilization is disastrously impacted? We are already adapting to the effects through efforts to protect coastlines, improve disaster responses, and breed heat-resistant crops. Time will tell.

From motions deep within the terrestrial core to gravitational interactions with its natural satellite and its overall orientation in space, we have tried to treat the Earth as just another planet, but, of course, have failed. We know this world too intimately.

We know of its two unique characteristics: an ocean of surface water and the presence of life. The water helps filter carbon dioxide out of the atmosphere. This process makes the Earth a more suitable environment for the evolution of life: Plants absorb carbon dioxide and emit oxygen whereas animals breathe the oxygen and release carbon dioxide. In addition, certain bacteria convert the atmospheric nitrogen gas into compounds that in turn are used in building the amino acids and deoxyribonucleic acid [DNA] found in plants and animals.

Of course, the singularly most unusual thing about the Earth is the presence of life at all, beginning some 4 billion years ago. Biologists have identified several million species of life on the Earth with millions more likely awaiting discovery (Figure 7.19).

Life, as we know it, depends on water: not just as a medium but because atmospheric water along with carbon dioxide gas moderates the terrestrial climate through the greenhouse effect, while the large heat capacity of surface water helps stabilize the climate. Seventy percent of the Earth's surface is covered with liquid water. Ninety-seven percent of the Earth's water is in oceans averaging 3.7-kilometers (2.3-miles) deep. Venus is too hot; a layer of water would evaporate. Farther from the Sun, Mars is too cold; what little surface water it has remains in a solid state. But as in the story of Goldilocks, the planet in the middle is 'just right'.

When looking for worlds of ice and fire, we need not look further than our Earth. While in the past these phenomena have occurred on a global scale ➤ Chapter 1 ➤, volcanoes and ice caps are

FIGURE 7.19   Fossilized coral *stromatoporoid* reef, the remains of one of the oldest life forms on the Earth.

local examples on present-day Earth. Life evolved here among these challenges and thrives in the most extreme terrestrial environments. ☞ Chapter 17 ☞ As we take a closer look at the worlds of the outer Solar System, we may need to refine our definition of 'just right' in order to expand our search for life beyond our current rocky outpost.

NOTES

1 With Dutch chemist Paul Crutzen, Molina and Rowland were awarded the 1995 Nobel Prize in Chemistry for their pioneering work on atmospheric ozone.
2 Approximately 90° to either side of the line connecting the centers of the Earth and Moon.
3 In the very distant past, at least two Snowball Earth episodes occurred when the Earth was completely frozen. ☜ Chapter 1 ☜.
4 For which they shared the 2007 Nobel Peace Prize.
5 Often referred to as global warming.
6 Such as solar, wind, and geothermal power.

8 Rusty Mars

Mars has long been considered a 'special' planet. For recent generations, astronomers and writers alike have imagined it to be the one most likely to harbor life beyond the Earth. Of the astronomers, Percival Lowell ⟨1855–1916⟩ is perhaps the best known for popularizing martian life.[1] Science fiction set on Mars begins with Percy Greg's *Across the Zodiac* in 1880 and includes Orson Welles's 1938 radio adaptation of *The War of the Worlds* and Matt Damon's 2015 appearance in *The Martian*.[2]

The first person known to track the complex pattern made by a bright red light in the sky lived in Egypt 4,000 years ago. Mars generally wanders eastward against the background stars but about every 26 months, it makes a westward loop or zigzag for 72 days. During this time, it is at its brightest, visible throughout the night. An Egyptian tomb painting, from the mid-second millennium BCE, records a conjunction of the five planets visible to the naked eye. Because Mars was moving backwards, or retrograde, at the time, its celestial boat is depicted as empty. After its retrograde excursion, Mars begins to set earlier until a year later it disappears in the glare of sunset for a few weeks before it re-appears rising just before sunrise. By the 5th century BCE, Babylonians were using periods of 47 and 79 days to predict the disappearance of the body they associated with Nergal, their god of war, plague, and death. Subsequently, it became associated with the classical war gods, Ares (Greek) and Mars (Roman) (Figure 8.1).

Just before the beginning of the telescope age, Tycho Brahe ⟨1546–1601⟩, an aristocratic Danish astronomer, compiled positions of Mars and other planets for 20 years. Working from these detailed records enabled Kepler to formulate his famous laws of planetary motion.

Mars averages 230 million kilometers (140 million miles) from the Sun. This is 1.5 au. At this distance, sunlight requires 12 minutes, 40 seconds to reach the planet.

Although Galileo is the first person to report seeing the *disk* of Mars through a telescope, Dutch physicist Christiaan Huygens ⟨1629–1695⟩ recorded the initial surface feature, a spot, in 1659. Once martian details could be seen, its characteristics seemed to resemble superficially those of the Earth. By timing the period between markings appearing and then reappearing, Italian-born astronomer Giovanni Cassini[3] ⟨1625–1712⟩ first measured the length of the martian rotation

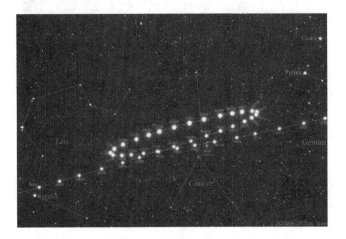

FIGURE 8.1 Mars retrogrades in the sky.

FIGURE 8.2   Mars as imaged by the Viking mission.

period; it is 24 hours, 40 minutes, which is close to the terrestrial 24-hour day. This interval now is called a sol. Similar days is quite a coincidence considering that the periods of the other terrestrial planets are measured in months while those of the Jovian planets are mere hours! (Figure 8.2)

   With better telescopes, white spots resolved into seasonally varying polar ice caps slightly offset from the top and bottom of the disk. From these, Herschel measured the angle between the martian equator and the ecliptic plane to be 30°. Therefore, he speculated that the martian seasons are like terrestrial ones. The modern value for that angle, or the obliquity of Mars, is about 25°. Here again,

FIGURE 8.3   Northern polar ice cap of Mars.

the amount of inclinations varies drastically across the Solar System, from nearly 0° to more than 90° (Figure 8.3).

On Mars, the northern polar ice caps shrink in the warmer summer when its northern hemisphere leans towards the Sun. It grows again during the colder winter that comes when the north pole is oriented away from the Sun. The same thing happens in the South. Mars and the Earth are the only solar-system planets to have polar ice caps, which wax and wane with the approach of winter and summer.

So, the seasons on Mars vary like those on the Earth. However, they last twice as long because the martian revolution period about the Sun (*its* year) is 1 year, 320 days. The martian year is longer because it is farther from the Sun than the Earth. The seasons differ in length from one another more so than they do on the Earth because Mars's orbit has greater eccentricity than does that of the Earth. For instance, northern-hemisphere Spring lasts for 194 sols, while Fall lasts for only 142 sols.

The family resemblance between Mars and the Earth does not, at first glance, confine itself to the 'numbers.' The martian surface area is roughly equal to that of the dry land on the Earth. The surface features of Mars look like those of the Earth, with thin white clouds floating above them adding yet more familiarity. In the 19th century, some astronomers were willing to go further: Through the eyepieces of their telescopes they thought that they saw green, seasonal vegetation spread in a hemisphere's spring and recede in that hemisphere's autumn. At the turn of the 20th century, Lowell was promoting the 'canals' he imagined he observed on Mars and the supposed intelligent creatures that built them[4] (Figure 8.4).

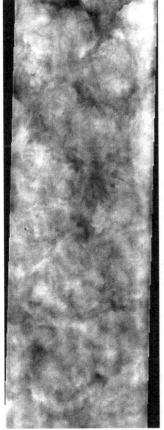

FIGURE 8.4 Carbon dioxide clouds on Mars imaged by Mars Odyssey.

FIGURE 8.5 Deimos as it appeared to Mars Odyssey.

Less fancifully, American astronomer Asaph Hall III ⟨1829–1907⟩ discovered two satellites orbiting Mars in 1877. The larger, inner moon is Phobos (Fear) and the smaller, outer moon is Deimos (Terror); the satellites are named for sons of Ares who accompany him in battle. Like the Earth's Moon, they revolve about their planet in the same time that they take to rotate; the same side is always facing Mars. So, yes, in that they orbit a planet, they are moons, but, otherwise, they are very different from the Moon. Phobos skims above Mars three times a day at a height of about 6,000 kilometers (3,700 miles), while Deimos takes slightly more than a day to orbit Mars at about 23,000 kilometers (15,000 miles) above it, roughly 6% of the Earth-Moon distance. No other satellites in the Solar System travel so close to their primary. Undoubtedly, a mission to either would find chunks of Mars in the form of meteorites. While both are cratered—Phobos boasts a crater nearly half its width—they contain too little material for their internal gravitational attraction to shape them into round worlds. Phobos also has grooves that may be the result of tidal stretching. The mean radius of Phobos is 11 kilometers (6.8 miles) whereas Deimos's is only 6.2 kilometers (3.9 miles). In fact, they look more like captured asteroids ☞ Chapter 9 ☞ than the Moon. Although today Phobos and Deimos can be seen every-so-often eclipsing the Sun from the martian surface, their sizes and distances prevent them from completely covering the solar disk. They are among the tiniest satellites in the Solar System (Figure 8.5). Small as they are, their situation cannot last. Phobos is literally inching towards the martian surface and cracking up. In less than 50 million years, Mars may lose a moon but gain a ring system.

The USSR and USA targeted Mars with space probes early in the space age. However, the first six missions failed. Mariner 4, one of a series of ten US missions to explore the terrestrial planets, finally flew by Mars on 14–15 July 1965, returning the first photographs of another planet.

In 1976, the USA placed the first landers on the martian terrain: Viking 1 and Viking 2. These missions and subsequent ones investigated the martian weather: Day-to-day there is a light breeze, coming from one direction in the morning and reversing in the evening. But it can get pretty windy at the landing sites occasionally, too: gusts of 100 kilometers per hour (60 miles per hour). On Earth, such speeds would be rated as 'storm force,' but, so thin is the martian atmosphere, an astronaut would barely notice them.

The Viking landers determined that Mars's surface is a porous regolith of mixed rock and dust. They helped verify that Mars actually is the 'Red Planet' because of iron oxide (rust) on its surface particles, and generally had a look around at sand dunes and vast boulder fields, some with morning frost.

The apparent Earth-Mars kinship actually suffered a great blow once scientists were no longer bound to examine Mars from a distance, when Earthlings could visit Mars—at least vicariously—using robotic space probes that returned objective data. It depends on what you count as success, but more than 30 missions have observed the Red Planet. Currently, 8 orbiters, 2 landers, 2 rovers, and a helicopter make their home at Mars.

FIGURE 8.6 Sojourner rolls off its lander and begins exploration of Mars.

The rovers are particularly interesting. Starting with Sojourner[5] on NASA's Mars Pathfinder Mission in 1997 and followed by twins Opportunity and Spirit in 2004, these remotely controlled vehicles allow microscopic investigation of the martian crust, complete with small grinders to investigate beneath the surface of rocks, and its atmosphere. Currently, the car-sized Curiosity, which landed in 2012, is performing investigations on Mars (Figure 8.6).

Mars 2020 joined this tradition when it landed in February 2021. As with almost all missions these days, it is an international collaboration with contributions from France, Spain, and Norway. Its components, the Perseverance rover and Ingenuity aircraft, study martian geology to assess its past habitability and signs of ancient life.

Perseverance is as big as a mid-sized car, weighs about a ton, and carries seven scientific instruments. It gathers rocks and soil samples for possible return to the Earth by a future mission. Rovers have ranged the surface of Mars continuously since 2004. As the first helicopter on Mars, Ingenuity will primarily demonstrate this technology for future use.

What have we learned from all of this effort?

First, those clouds? Most are not made of water, like terrestrial clouds; they are carbon dioxide ice flakes. In fact, the entire atmosphere of Mars is made utmost of this soda fountain gas, not the Earth's nitrogen and oxygen. It is far too thin to breathe, anyway.

Ice caps? Again, they are not solely frozen water as on the Earth, but more carbon dioxide in a solid phase—dry ice. The seasonal changes of the ice caps are due to the deposition and sublimation of carbon dioxide, which lies on top of residual or permanent ice caps of water. Liquid water is not visible on the surface of arid Mars, though the weight of all that ice may create a thin

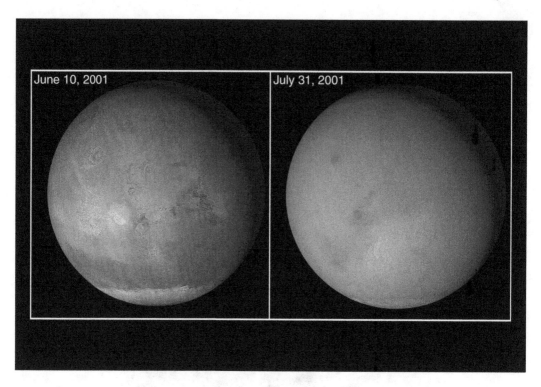

FIGURE 8.7   Mars Global Surveyor image showing a major dust storm on Mars.

layer of liquid water underneath. Scientists are learning more about liquid water on Mars. NASA's Mars Global Surveyor Mission, which arrived in 1997, noted that underground water may be seeping out of the walls of craters, creating small gullies.

And as for that imagined 'greenery,'[6] it was actually huge seasonal dust storms over a desert planet that last for months. Such storms travel 100 kilometers per hour (60 miles per hour), are the size of Earth continents, and are so thick and dark that they obscure 99% of the sunlight from reaching the surface! (Figure 8.7)

The thin atmosphere is perhaps the most confounding fact to those ready to set up camp on Mars. It is more rarified than a good laboratory vacuum here on the Earth. And that is just the troposphere; Mars has no stratosphere. The average *surface* pressure is 50 times less than that at the top of Mount Everest. The exact pressure varies with season: higher in the summer when carbon-dioxide ice sublimates from the poles, lower in the winter when the carbon dioxide snows back onto the poles.

A thin atmosphere allows things easily to come and go. Meteoroids from space collide nearly unhindered with Mars and leave a cratered surface. ☞ Chapter 9 ☞ For instance, Hellas Planitia (Hellas[7] Plain) is 7.2 kilometers (4.4 miles) deep and 2,300 kilometers (1,400 miles) in diameter.

Rampart craters like Tooting[8] are found primarily on Mars. On their sides are what look like mud slides, which beg the question of subsurface water (Figure 8.8).

The thin martian atmosphere also allows daytime heat from the more-distant Sun to leave Mars quickly at night. Temperatures on Mars range from −125°C (−195°F) at the poles in winter to 20°C (70°F) on a nice summer's day near the equator. The greenhouse effect is slight on Mars. Most of the martian carbon dioxide is entombed in its rocks, trapped due to the lack of geologic activity with which to transport it once again to the surface. (See below.)

Worse, a thin atmosphere lets in harmful radiation from space—a serious detriment to hypothetical martian life. With only a trace amount of oxygen, an ozone layer cannot form to block ultraviolet light from the Sun.

FIGURE 8.8   Mars Odyssey image of rampart crater Tooting on Mars.

Yet, an astronaut in a space suit *could* survive on Mars. The space suit, with its requisite heater, oxygen supply, radiation shielding, *etc.*, will not be an encumbrance because, due primarily to its lower mass, Mars has a surface gravity only 37.5% that of the Earth. Once suited up, she would have amazing views in every direction.

Looking up, she would see a *pink* sky, colored by dust particles inserted into the atmosphere by 'dust devils.' Below, she can take her pick: The southern hemisphere is older and more cratered than the northern hemisphere, which is a younger volcanic plain (perhaps due to a mega-impact there early in Mars's history). ☛ Chapter 9 ☛ Alternatively, in the North, Tharsis Montes (Tharsis[9] Mountains) is a series of great shield volcanoes rising above the martian plain. Not versions of the 'little' ones we see on the Earth, martian volcanoes are enormous. The leader of the pack, Olympus Mons,[10] is 25 kilometers (14 miles) high and has a caldera bigger than the American state of Rhode Island. With an overall diameter of nearly 620 kilometers (370 miles)—its base is bigger than the state of Arizona—it is the largest volcano in the Solar System. Olympus Mons does not seem to be erupting right now; it may be extinct (Figure 8.9).

The huge size of martian volcanoes is evidence that Mars is no longer geologically active in the form of plate tectonics like the Earth is. Instead of a chain of smaller volcanoes, like the Hawaiian Islands, created as a crustal plate moves over a hot spot, martian volcanoes are isolated but colossal. Therefore, the associated hot spot must remain under one place and create a single volcano that keeps getting bigger and bigger.

The martian volcanoes are probably all extinct. So, most of the once fluid interior of Mars may no longer be liquid at all. The martian crust behaves like one, single plate with no churning mantle to propel plate tectonics.

However, planetary scientists expect Mars's interior to be generally similar to that of the Earth: crust, mantle, and a partially liquid core of molten iron. The next aim of Mars exploration is to return rocks from its crust. Amazingly, geologists already have collected about a dozen such rocks in the form of meteorites that traveled freely to the Earth after being ejected from Mars by past violent impacts. ☛ Chapter 9 ☛

Only the bigger terrestrial planets still have hot interiors because the volume [V] of a spherical planet increases as the cube of the radius [R]

$$V = (4/3 \cdot \pi \cdot R^3$$

whereas its surface area [A] increases only as the square of the radius

$$A = 4 \cdot \pi \cdot R^2$$

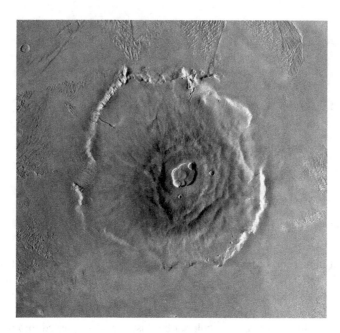

FIGURE 8.9   Viking image of the martian volcano Olympus Mons.

As its size increases, a planet gains volume more quickly than surface area. Hence the volume-to-surface area for a small planet is less than that of a large planet. With more volume and less surface area, a larger planet has more mass to store internal heat, but less area through which to radiate this heat away into space. (Thermos ® bottles always have small openings, to reduce the area through which heat can escape.) If the densities are the same, the smaller planet will cool down more quickly than the larger planet.

We experience this effect in our everyday world. Bake two potatoes, one small and one large, to the same temperature. Remove them from the oven and let them cool. Which one cools faster? The small one, of course!

Stored internal heat keeps rock liquid and allows for volcanism and plate tectonics. Mercury and Mars likely have radiated away most of their internal heat, while Venus and the Earth still retain a considerable amount of it. In addition to residual heat from their formation, the decay of radioactive elements also adds heat to planetary interiors, and the more planet, the more radioactive elements.

Recall that seismometers on the Earth and the Moon record vibrations that describe the subsurface conditions through which they are passing. The terrestrial crust dampens earthquake vibrations in a few minutes while a moonquake can ring the Moon with its drier crust like a bell for up to 10 minutes. What about Mars?

While the Viking landers carried seismometers, they failed to capture any definitive seismic activity in 1976. The first instrument failed to deploy while the second was not in direct contact with the martian surface, so it returned a wealth of meteorological data instead.

Beginning in 2018, the Interior Exploration using the Seismic Investigations, Geodesy and Heat Transport [InSight] lander began studying marsquakes to understand the structure of Mars better. The signature pattern of marsquakes appears to be somewhere between the two examples of the Earth and Moon.

Returning to our tour of Mars, Valles Marineris (Valley of the Mariners[11]) is a gorge up to 10 kilometers (6 miles) deep, more than 200 kilometers (130 miles) in width, and a fifth of the martian circumference in length. But never mind the numbers, Valles Marineris makes the Grand

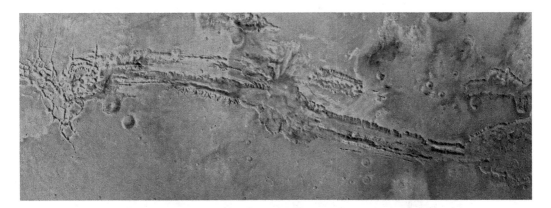

FIGURE 8.10 Valles Marineris as imaged by the Viking Mission.

Canyon look like a ditch! The Colorado River eroded the Grand Canyon out of the Colorado Plateau. While some eastern channels in the vast Valles Marineris system may have been formed by water, it probably began as fissures in the cooling martian crust that have been further eroded. The crust additionally may have been deformed by the rising lava fields of the adjacent Tharsis Planitia and its three associated volcanoes (Figure 8.10).

Mars has other valleys, of course. Some are ancient lava flows. However, others look suspiciously like river valleys on the Earth, more specifically, dry gulches or arroyos. They have sand bar islands and river deltas—just like those in the Southwest USA (Figures 8.11 and 8.12). In addition, more than one dry lakebed graces the martian surface along with evidence of a once-upon-a-time large northern ocean.

Moreover, Curiosity investigated stones that appear to have been smoothed by running water. It also explored a region with mineral salt deposits, which may be the remains of a briny lake; salt settles out of the ocean on the Earth. In addition, Curiosity sampled clay in several areas, including a large deposit at Aeolis Mons (Mount Aeolis[12]); clays usually form in the presence of water. A flood of Biblical proportions appears once to have taken place in the Ares Vallis (Ares Valley), which NASA chose as the Mars Pathfinder landing site.

Such discoveries are intriguing because they suggest that at one time liquid water flowed on the martian surface—20% of the planet once may have been covered with it. And water is almost

FIGURE 8.11 Sand bar presumably formed by flowing water on Mars, as imaged by the Mars Reconnaissance Orbiter [MRO].

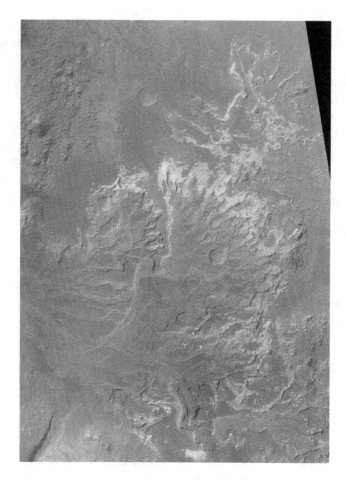

FIGURE 8.12 Dry river delta as imaged by the Mars Global Surveyor.

certainly a prerequisite for life. Mobile robot explorers have gone on to detect the presence of dry rock and minerals on Mars that must have formed in the presence of water. Where, then, did all this water go?

Some of the water is hidden under the northern carbon-dioxide polar cap and is exposed as ice in the summer. Much of the rest of it is either stored as permafrost, starting less than a meter under the martian surface, or bound together with other molecules as mineral hydrates. The bottom of at least one deep crater is lined with ice. Occasionally, drips of briny water melting from icy cracks flow down hillsides and crater walls. Mars geologists even think that they have found sand-filled underground 'lakes' up to 19 kilometers (12 miles) across. Still, a problem remains: If a future astronaut were to pour out a glass of water onto Mars, it would boil away! (Figure 8.13)

If you read the label on a box of cake mix or other baked goods carefully, you will see special instructions for people who live at high elevations like Denver, Colorado, or Davos, Switzerland, where the air is thinner. The temperature at which water boils decreases with atmospheric pressure. The atmospheric pressure at the surface of Mars is so low that liquid water would boil away immediately. This raises another question: Where, then, did the atmosphere, which must have been denser in the martian past, go?

The atmosphere may be hiding in plain sight. The polar caps, snow clouds, and frost on the martian ground all are made of carbon dioxide, the principal ingredient of the martian atmosphere.

FIGURE 8.13 Martian gullies as imaged by the MRO. Were they produced by flowing water?.

Mars, at one time, may have become so cold that its atmosphere froze out; Mars now may be experiencing some sort of ice age.

What could change martian climate so dramatically? As noticed by Herschel, the comparable obliquities of Mars and the Earth imply similar seasons. The extremes of a planet's seasons have to do with its obliquity. Maybe Mars once had a more extreme tilt, complete with winters that solidified its atmosphere.

Not only do planets interact gravitationally with the Sun and their moons, but planets also interact with each other, although to a lesser extent. Modeling of these interactions indicates that the obliquity of Mars varies from 0° to 60°; its current 25° orientation is a rare, mild period. While, as on the Earth, most martian rock is igneous (solidified lava), a clear pattern exists of sedimentary rock layers, made in a watery environment, that could have been caused by such changes in its axial tilt. This is our best testimony yet that Mars's global climate has changed. While speculative, evidence of obliquity-driven climate change for Mars is an intriguing possibility (Figure 8.14).

The terrestrial climate does not change as fast or as greatly as the martian; the axis of the Earth is stabilized by its Moon. Unfortunately for Mars, it has no massive satellite to lend its protection.

Nor does Mars any longer have a strong magnetic field to protect it from the solar wind. Old terrain exhibits magnetized material of alternating polarity, likely preserving remnants of an ancient dipole field some 4 billion years ago. So, charged particles carried by the solar wind can crash into the molecules of the martian atmosphere. The impact can send the molecule away from the planet. This slow particle-by-particle removal of the atmosphere is called sputtering. In 2017, results from the Mars and Volatile Evolution Mission [MAVEN] in orbit around the planet, and Curiosity rover on the surface, indicate that sputtering is affecting the martian atmosphere now and probably had a greater effect in the past when the young Sun produced more intense solar wind. Thus, Mars is losing its atmosphere permanently.

Clearly, though, Mars was more Earth-like in the distant past, and that nagging question returns: Under such milder conditions did life evolve there—perhaps even before that on the Earth? Martian life may be extinct now—its traces left only in fossils. Yet, if life arose on Mars, it heralds a paradigm-shifting answer to the question of how many 'living' planets are out there. Right now, the answer is only one.

What Viking 1 and 2 revealed about martian weather and scenery was all well and good, but their principal mission was to search for signs of life. Nobody really expected a martian creature to run before their cameras. Nevertheless, what about simpler life, for instance, bacteria? Are they protected from incoming radiation by the martian soil? To this end, the Vikings had a mechanical arm that dug samples of martian dirt and placed them in a special Labeled Release Experiment [LRE] box aboard the craft. This chamber contained nutrients on which biologists hoped martian microbes might gorge themselves. Then, as all life forms eventually must after a good meal, the microbes would—to put it delicately—change the chemistry of the chamber. The instrument could

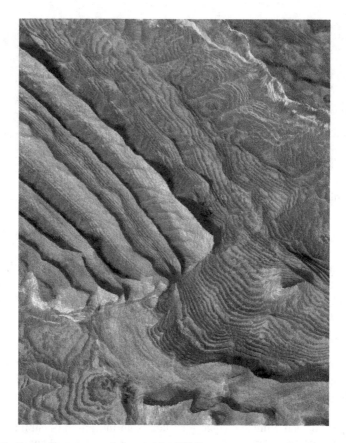

FIGURE 8.14   Martian sedimentary rock imaged by MRO.

measure such a chemical change and relay the results to the awaiting scientists back on the Earth (Figure 8.15).

Ideally, an experiment would have a 'yes' or 'no' answer. However, the Universe does not always deliver. 'Yes,' the experiments detected chemical changes in samples aboard Viking. 'No,' the LRE detected nothing that also could not be explained by non-living chemistry. The results were inconclusive.

Later robots on the surface revealed complex organic molecules on Mars. These are the building blocks of life, but do they indicate anything about life on *Mars*?

A complimentary debate rages over the presence of methane in Mars's atmosphere. Some ground-based telescopes and space probes measure a significant amount, others do not. While a trace gas, methane is important because it does not remain intact within the atmosphere very long. It must be replenished. Natural abiotic processes can do this, but on the Earth, an important source of methane is life. ☞ Chapter 17 ☞

Where does this leave us? While the case for life on Mars does not appear promising, proving a negative is difficult. Life on the Earth thrives in very inhospitable places, such as Antarctica. What is more, planetary scientists keep making discoveries that remove impediments to life on Mars. For example, Curiosity rover detected boron, a life-building element, in Gale Crater in 2017. To this day, the question of a living Mars goes unanswered.

Mars was more Earth-like in the past. Now, it is not. Could it be made so again? We use the provocative word 'made' on purpose. The process is called terraforming. It was the subject of a classic Arnold Schwarzenegger movie,[13] but is only just beyond our current technological capability. One idea would be to seed some genetically modified plants designed to withstand the

FIGURE 8.15   Viking lander prepares to use its robotic arm to dig into martian soil.

harsh martian environment. These would release oxygen, building up the martian atmosphere and possibly forming an ozone layer to shield its surface from ultraviolet radiation. Another would spread dust on the polar caps to vaporize the carbon dioxide trapped there because a dark surface retains more solar heat. Either way, Mars would grow warmer, its atmosphere enriched in oxygen would become thicker, water would flow once again, and, eventually (after a thousand years or so!), we could colonize. ☛ Chapter 17 ☛

An ethical question remains: All this would lead to the demise of any indigenous martian life or martian life to come. Do you think we would do it, anyway? We believe that we all know the answer to this question already. Can you recall a place here on the Earth where colonists resisted the urge to change things on their own behalf? (Figure 8.16)

FIGURE 8.16   Martian sunset regarded by the Mars Curiosity Rover.

NOTES

1 More successfully, Lowell initiated the search for a planet beyond Neptune that led to the discovery of Pluto. ☞ Chapter 14 ☞.
2 In which he plays Mark Watney for the film adaptation of the 2011 novel by Andy Weir.
3 Best known for identifying the Cassini Division in the rings of Saturn.
4 The human brain is quite capable of connecting unrelated features seen by eye and turning them into visions of coherent structures.
5 Named after Sojourner Truth ⟨circa 1797–1883⟩, a former enslaved woman and African-American civil rights activist.
6 The human brain easily can misinterpret hues at the limit of the eye's ability to resolve color.
7 Using the Greek name for Greece.
8 Named for a London suburb.
9 After a city far to the west mentioned in the Bible.
10 After Mount Olympus in Greece, the traditional home of their gods.
11 After Mariner 4, Mariner 6 and Mariner 7, which flew by Mars in 1969. Then, Mariner 9 mapped the planet extensively in 1971–1972.
12 Named for the region of ancient Greek settlements along the Western Coast of Anatolia, although the Curiosity mission team prefers to call it Mount Sharp after American geologist Robert Sharp ⟨1911–2004⟩.
13 *Total Recall* (1990).

9 Small Bodies: Asteroids and Meteoroids

Imagine now that it is a lovely Spring afternoon. You have the grass cut just so. You are ready to settle back in your hammock and watch a match on TV. You have not a care in the world.

Suddenly, an errant space rock tens of kilometers across tears into the atmosphere above your head at something like 17 kilometers per second (11 miles per second). Atmospheric friction hardly slows it, though, in the instant it takes to reach the ground. It does not stop there. It plunges 20 kilometers (12 miles) into the Earth's crust and explodes with an energy equivalent greater than that of the nuclear arsenals of the human race. The blast vaporizes the people, trees, houses, and very rock of what was once your county. The eruption carves a crater 200 kilometers (120 miles) across in the remaining landscape. It throws pulverized rock and dust upward into space. Some rocks fall back to Earth becoming fireballs themselves heating the atmosphere and triggering wildfires. However, enough dust remains in the upper atmosphere to block sunlight for weeks, months, even producing a cold, artificial twilight. The initial heat enables large quantities of atmospheric nitrogen and oxygen to combine chemically forming a smog of nitrogen oxides. The nitrogen oxides will go on to combine with atmospheric water to form nitric acid, the result being a worldwide rain nearly the strength of sulfuric acid that will wash the remaining dust from the atmosphere. When all the debris finishes settling back to the terrestrial surface, it will form a distinctive, worldwide layer centimeters thick. Unfortunately, neither you nor any creature that saw the crash survived making lawn mowing and the outcome of the prospective match irrelevant (Figure 9.1).

Is this story the trailer for a science-fiction disaster movie? Or the rattling of a pessimist fixated on the worst-case scenario ever? Or something that should keep us awake at night? The answer lies somewhere in between. On one hand, this nightmare scenario is based on computer simulations and probably accurately describes what could happen when a large space rock, or more accurately a meteoroid or comet ☛ Chapter 13 ☛ encounters the Earth. It is close to the leading theory for the event that triggered the Cretaceous–Paleogene [K-Pg[1]] extinction event 66 million years ago, which ended the 180-million-year reign of the dinosaurs. On the other hand, such an encounter happens perhaps once every 100 million years or so. Nobody has ever witnessed such an event. Nonesuch has occurred in historical times. Nonetheless, in a book that routinely deals with millions and billions of years, we should explore what might happen over geologic time.

A million kilograms (1,100 tons) of interplanetary material arrive at the Earth daily in the form of interplanetary dust and meteoroids. Interplanetary dust grains inhabit the plane of the Solar System. Traveling much faster than a speeding bullet, they can be dangerous to spacewalking astronauts' suits and even spaceship walls and windows. In 1993, ESA discontinued its Olympus telecommunications satellite after a suspected micrometeoroid strike (Figure 9.2).

The largest grains of interplanetary dust, or micrometeoroids, continually intersect the Earth. They are still so lightweight that, once in the atmosphere, they float gently down to the ground. They do not cause any trouble, but there are a lot of them (Figure 9.3).

In addition to these barely discernable pellets, other small rocky and metallic bodies called meteoroids routinely strike the Earth. Meteoroids have random orbits about the Solar System. Some may be the crumbled result of bumps between larger meteoroids or 'chips off the old block.' Others, traveling in clusters, are the spent remains of comets, otherwise icy bodies associated with the outer regions of the Solar System. ☛ Chapter 13 ☛ Meteoroids range in size from about a grain of sand, such as the micrometeoroids, to about 1 meter (40 inches) in diameter. Bigger than this, astronomers begin to refer to these space rocks as asteroids. To date,

FIGURE 9.1   Artist depicts an impact that would effect life on the Earth forever.

FIGURE 9.2   Interplanetary dust is visible in a dark sky as zodiacal light.

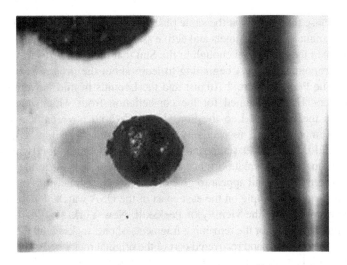

FIGURE 9.3 Micrometeorite. These typically range in size from 0.1 to 1.0 millimeter (0.004 to 0.04 inch).

2015 C25[2] is the smallest asteroid observed from Earth with a diameter of about 2 meters (80 inches) as determined by radar in 2015.

Meteoroids cause the smaller craters that appear abundant on planetary surfaces with little or no atmosphere. Moving at supersonic speeds, meteoroids explode upon impact, excavating a crater. Creating a large basin requires an asteroid or large comet. ☞ Chapter 13 ☞

In the case of the Earth, though, heat builds up as the meteoroid plunges through the mesosphere, compressing air extremely quickly. Meteoroids burn, and their fiery destruction produces those characteristic bright streaks in the night sky called 'falling stars,' 'shooting stars,' or more properly—because they have nothing to do with stars—meteors. At a dark site on a clear night, you ordinarily might be able to see ten meteors per hour, more in the hours before dawn in the Fall and fewer on a Spring evening. Alas, lunar colonists would see no meteors because the Moon lacks an atmosphere; they would see an impact flash when the meteoroid strikes the ground (Figure 9.4).

When the Earth intercepts a belt of meteoroids, the number of meteors you can see on a given

FIGURE 9.4 Bright meteor (bolide).

night increases. Because the Earth is at the same place in its orbit at the same time each year, these meteor showers are annual affairs. Comets and active asteroids can create such belts in their wakes. Comets lose dust when they are close enough to the Sun that their ice turns to vapor creating their legendary tails. Astronomers are just beginning to learn about the processes by which active asteroids eject dust. The Perseids in mid-August and the Leonids in mid-November are two of the better-known showers. Each is named for the constellation from which its meteors appear to radiate, i.e., Perseus the hero and Leo the lion. Although both have cometary origins, Comets Swift-Tuttle[3] and Tempel-Tuttle,[4] respectively, active asteroid Phaethon[5] creates the Geminid meteor shower in mid-December that appears to radiate from Gemini, the twins (Figure 9.5).

Occasional, large meteoroids create bright meteors called bolides. A bolide may be visible for several seconds, change colors, and appear to give off smoke. It is sometimes called a 'fireball.' On 9 October 1992, thousands of people on the east coast of the USA watched a greenish bolide travel northeast to crash eventually in the vicinity of Peekskill, New York.

Sometimes a large meteoroid, or the remaining fragments of one, makes it all the way to the ground. In the Peekskill case, Michelle Knapp recovered part of the original rocky body under her 80 Chevrolet Malibu through which it had smashed, missing the gas tank; a rare instance of damage increasing the value of a vehicle. A generation earlier, in 1938, another meteoroid crashed through the roof of Ed McCain's garage in Illinois, through the roof of his car, and embedded itself in the thankfully empty passenger seat. You can see the rock and its resting place, both preserved in Chicago's Field Museum.

Because traveling through the atmosphere slows a meteoroid, these small objects do not make large craters, which is fortunate for Ann Hodges ⟨1920–1972⟩. One smashed through her ceiling in 1954, rudely disturbed her nap, and bruised her.[6] Instead, they usually join the billions of other rocks strewn about the ground. Like the micrometeorites mentioned earlier, successfully traversing the atmosphere changes a meteoroid into a meteorite, at least in terms of vocabulary.

Many meteorites seem to have come from differentiated bodies. They are usually made of rock or metal although a small population that consists of both rock and metal is known. Until the 20th century, most meteorites collected were largely made of dense metal even though rocky meteorites are more common. A dark, unusually heavy-to-lift object lying on the ground garners more attention than a rocky meteoroid, which may be overlooked as just one more terrestrial rock. The iron dagger buried with Tutankhamun ⟨circa 1341–1324 BCE⟩. a pharaoh of Egypt's 18th dynasty, appears to have been made from a meteorite. Antarctica is a good place to look for meteorites because of their contrasting appearance against the blue ice and because the movements of the ice sheets tend to concentrate meteorites in specific areas.

FIGURE 9.5 Meteor shower. This is a long photographic exposure; one is unlikely to see these many meteors at once.

Meteorites are exciting because they are one of the few astronomical objects you can hold in your hand, rather than just look at through a telescope or via a space mission. When cut open, metal meteorites display a pattern of large crystals, which is just what a molten object that cooled very slowly in the depths of space should look like. This characteristic is known as the Widmanstätten pattern (Figure 9.6).

Be careful: These three words—meteoroid, meteor, and meteorite—sound similar but mean different things. A meteoroid is the physical object that produces the optical phenomenon we call a meteor. Only a few meteors survive the trip through the atmosphere to become meteorites, or rock specimens. To be more confusing, these words share a root with meteorology, the study of weather, because they are derived from *meteoros*, a Greek expression, meaning lofty or high in the air.

If all these things dropping out of the sky are mostly harmless, what could create the terrifying scenario with which we opened this chapter? The surfaces of the other terrestrial planets are scarred by large impact craters. Yet, only in 1960 did American geologist Eugene Shoemaker[7] ⟨1928–1997⟩ convincingly identify the first *bone fide* terrestrial impact crater: Barringer Crater; it is a 49,000-year-old crater in the dry, preserving highlands of Arizona (just off US Interstate 40). However, this particular hole in the ground is now 1.2 kilometers (0.75 miles) across and 170 meters (550 feet) deep. Geologists estimate that the impactor had a diameter of 'only' about 40 meters (130 feet). While impressive to human eyes, it is trivial compared to the scars visible on other worlds (Figure 9.7).

Sometimes spotting an impact crater is difficult just because it is big. You may be standing right on top of one, but from your perspective, there is little to see. The German town of Nördlingen lies within a giant impact crater. Since its formation 15 million years ago, it has been modified by flowing rivers and eroded by wind and water. One can just barely make out a ring of hills in the distance if one knows to look for it—the crater rim. Another impact feature, Colônia Crater, hosts the town of Vargem Grande, Brazil, and was not known until 1960.

To get the right perspective, you need to look at a crater from overhead. Craters on the Moon are easily seen. We see them from above, and meteorological and geological processes have not eroded them away. Aerial photography helps here, but our first systematic study downward onto our planet only came about with the advent of orbiting satellites. Even then, there are problems of identification. Not only has the crater likely been eroded by flowing water, plate tectonics, and the like, but also life itself tends to cover and make it difficult to see an old impact crater (Figure 9.8).

FIGURE 9.6   Iron meteorite sliced to reveal its interior.

FIGURE 9.7   Barringer Crater.

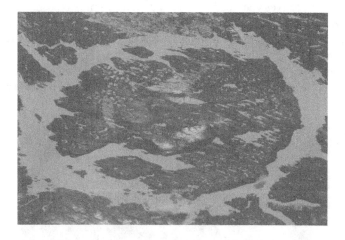

FIGURE 9.8   100-kilometer (60 mile) diameter Manicouagan Reservoir in Canada was formed 210 million years ago by the impact of a 5-kilometer (3-mile) diameter asteroid. It has since been silted in by rivers. This image was made by the crew of the International Space Station.

In addition, other processes can make circular features on the Earth: salt domes, sink holes, volcanoes... The proof lies in onsite inspection, the search for rock that has undergone exposure to extreme heat and pressure. Geologists have identified 190 impact craters[8] on the Earth. The largest is the Vredefort crater in South Africa, which is about 300 kilometers (190 miles) across and dated to be about 2 billion years old. In particular, the Chicxulub crater is associated with the K-Pg extinction, but this 180-kilometer (110-mile) diameter feature is buried beneath the Yucatán Peninsula in Mexico and the Gulf of Mexico. Recognized craters tend to lie in areas of low geologic activity and little human population—we just have not looked at the more remote sites on our globe. How many must lie beneath the ice caps and oceans that cover most of the Earth? The Earth is certainly not immune from devastating impacts (Figure 9.9).

A meteor may cause damage without reaching the Earth's surface or creating a crater. One of the great natural mysteries of modern times is the Tunguska Event, so named because scientists do not quite know what happened in central Siberia in 1908. Eyewitnesses reported a huge fireball, but none of the scientific expeditions have located a telltale crater. Instead, forests of trees were burned or knocked down radially in every direction from ground zero. Astronomers speculate that the Tunguska Event was the result of an asteroid or comet blowing up in the atmosphere before it hit the ground. You likely have not heard of the Tunguska Event because it happened in such a thinly populated area. Nevertheless, a similar event would be devastating if it occurs over a more inhabited locale, like a city. A smaller meteoroid exploded over Chelyabinsk, a Russian city, in 2013; the pressure wave it generated shattered glass windows all over town and sent more than a thousand injured people to the hospital (Figure 9.10).

Thinking about it, though, with so many smaller bodies 'out there,' some huge ones probably are orbiting the Sun still. However, many large meteoroids already terminated their orbits in the making of the great craters and basins on the Moon and planets during the Era of Heavy Bombardment. ⇥ Chapter 1 ⇥ Therefore, encounters between such bodies and planets should no longer occur frequently. While crashing high-mass bodies are now rare, even should the probability be one in a million years, that is of little comfort, if this happens to be that millionth year.

Although Mercury, Venus, Earth and Moon, and Mars are the five major terrestrial worlds of the Solar System, other, smaller bodies exist that fit this definition, at least insofar as that they are made of rock and metal. Most of these smaller worlds are in the gap between the orbits of Mars and Jupiter. These 'mini-planets,' if we do not want to call them full-fledged planets, are the asteroids. They were mentioned earlier as overgrown meteoroids (Figure 9.11).

On 1 January 1801, Piazzi noticed what he thought was a faint star. However, the following night, it had moved relative to the other stars indicating that it was not a star. Watching its changing position over many nights, he determined that it was a Solar System object in the gap between Mars and Jupiter. Piazzi thought he had discovered a missing planet, which he named Ceres.

As discussed earlier, while finding a previously unknown 'planet' was not surprising in 1801, the crowding of the region between Mars and Jupiter with these small bodies was unexpected.

FIGURE 9.9 Distribution of known terrestrial impact craters.

FIGURE 9.10   Tunguska site.

FIGURE 9.11   Ryugu, a typical small asteroid, imaged by Hayabusa 2.

Discovering multitudes where one planet might be expected led astronomers to classify these tiny worlds as asteroids rather than planets. ➥ Chapter 2 ➥ Most asteroids orbit the Sun at radii between 2.2 and 3.2 au in a ring now known as the Asteroid Belt[9] (Figure 9.12).

By now, almost *one million* such objects have been discovered, had their orbits plotted, and been named. Only 26 have diameters greater than 200 kilometers (120 miles), and this number is unlikely to rise. Astronomers probably have discovered 99% of those with diameters 100 kilometers (120 miles) and greater. Ten to 100 kilometers (12 miles to 120 miles)? Half. However, modern imaging techniques increase the quantity of known smaller asteroids daily. For all those

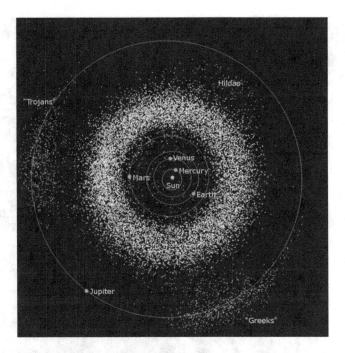

FIGURE 9.12 Diagram of the Asteroid Belt, including Hildas, and Trojans.

found so far, we can reasonably assume that *millions* or more exist that are too small or distant to have been detected yet.

Although 'asteroid' means 'star-like,' they are not star-like at all. In reality, they behave like planets: They orbit the Sun, shine by reflecting sunlight, may host satellites, and so forth. 'Planetoid' might be better, but we are stuck with asteroid for historical reasons. Astronomers who specialize in these small bodies prefer calling them 'minor planets,' recognizing that a continuum of size and mass exists from interplanetary dust particles through meteoroids to asteroids and comets and, eventually, to planets. While we tend to think of asteroids as rocky bodies without tails and comets as icy bodies with tails, a continuum of composition from rocky to icy exists between these bodies as well. So, the term 'minor planet' has some merit although it has never caught on with the public (Figure 9.13).

Whatever you prefer to call them, 20th-century observers began to classify asteroids further by similarities in their orbits and similarities in their colors. Occasionally collisions among asteroids or with meteoroids can produce families of smaller asteroids with similar orbits and compositions. Astronomers have identified more than 100 smaller bodies that were once part of Pallas, possibly including Phaethon. Looser groups of asteroids with broadly similar orbits that may not share a clear common origin also exist. The Atens[10] and Apollos,[11] named for the first of their kinds, have orbits that cross the Earth's.

The color of an asteroid is an indication of its composition with three main types. Most common (75%) are the dark, grayish chondrite, or C-type, asteroids made of clay and other minerals known as silicates, which are based on compounds of silicon and oxygen. Then, about 17% of asteroids belong to the relatively bright greenish to reddish stony, or S-type, made of silicates plus nickel and iron. Finally, a small number of asteroids fall into the reddish metallic, or M-type, made primarily of nickel-iron. (These percentages may be biased because the dark C-types are harder to see.) Other types exist, such as E-type and Q-type, based on unique properties of their spectra. Early collisions among bigger, differentiated bodies in the Asteroid Belt that resulted in shattered chunks of their mantles and cores could explain these differences in composition.

FIGURE 9.13 Montage of asteroids.

Are all these chunks of rock and metal circling the Sun in a common belt the pieces remaining after the break-up of what was once a single planet? It would have been a very small one because, had all the asteroids ended up as a single planet, it would have had less mass than the Moon. This scenario is reminiscent of Superman's fictional Krypton, the planet that blew up. However, nobody has discovered a way to explode a planet, even a small planet. More likely the Asteroid Belt represents a planet that somehow was prevented from becoming whole in the first place. Maybe the gravity of massive Jupiter nearby had a disrupting influence.

Certainly, the gravitational influence of Jupiter shapes the Asteroid Belt today. In 1866, American astronomer Daniel Kirkwood ⟨1814–1895⟩ analyzed the orbits of the 87 asteroids then known. Within the Asteroid Belt, he found no asteroids orbiting the Sun at specific distances. Later studies revealed that any asteroids at those radii would put them in unstable resonances with Jupiter. The Kirkwood Gaps exist where an asteroid would orbit the Sun three times for every Jupiter orbit (3:1), five times for every two Jupiter orbits (5:2), seven times for every three Jupiter orbits (7:3), and twice for every Jupiter orbit (2:1). An asteroid that migrates into one of these regions experiences periodic gravitational pulls from Jupiter that change its orbit until it is removed from the resonance. Kirkwood also suggested that a similar process may shape the gaps in the rings of Saturn. Alternatively, Jupiter's gravitational influence can concentrate asteroids into distinct groups in stable resonances. One such group is the Hildas,[12] on the outer edge of the Asteroid Belt, revolving about the Sun three times for every two Jupiter orbits.

Sunlight also affects the orbits and spins of asteroids. The hot, daylit side of an asteroid radiates heat back into space, causing a small thrust on that side and a recoil acceleration due to Newton's Third Law. Over millions of years, the asteroid's orbital motion changes significantly, either speeding up and moving away from the Sun or slowing down and spiraling closer to the Sun.

Polish engineer Ivan Yarkovsky ⟨1844–1902⟩ first noted this reaction. The influence is tiny compared to gravitational forces and is most significant for small bodies, those about 10 centimeters to 10 kilometers (4 inches to 6 miles) in diameter. This Yarkovsky Effect is essential in predicting the possibility that Near-Earth Objects [NEOs] will impact the Earth.

The Yarkovsky, O'Keefe, Radzievskii, and Paddack [YORP] Effect is similar to that of Yarkovsky. Sunlight also exerts a rotational force (torque) on spinning asteroids with surface irregularities and albedo variations that increases their rotation rate. Therefore, small asteroids spin faster than large ones. If rotating fast enough, an asteroid will break apart. Consequently, the YORP Effect creates binary and tumbling asteroids. In addition, the YORP Effect modifies the obliquity of an asteroid changing its pole towards 0°, 90°, or 180° relative to the ecliptic. For instance, members of the Koronis family have obliquities near 0° or 180°, with nothing in between. Measuring the Yarkovsky and YORP Effects is a top objective of the Origins, Spectral Interpretation, Resource Identification, Security-Regolith Explorer [OSIRIS-REx].

For a close-up view of an asteroid, we must turn to the robot eyes of spacecraft. For instance, during its 1969 flyby of Mars, Mariner 7 collected one image of the martian satellite Phobos while Mariner 9 took images of Deimos as well. Recall that astronomers think they were once asteroid-belt bodies that somehow ventured too close to nearby Mars and were gravitationally captured into its orbit.

These small bodies proved typical in that they do not contain enough mass to press themselves into spheres. Asteroids come in a bewildering set of shapes. While some rotate about a well-defined axis, some asteroids tumble unpredictably through space (Figure 9.14).

Jupiter, beyond the Asteroid Belt, also has several irregular satellites, such as Amalthea,[13] that may be captured asteroids, too. ☛ Chapter 12 ☛ However, the neatly near-circular orbits of Phobos and Deimos indicate that their origins may be more complicated. When the Earth captures a passing asteroid, it orbits us in an eccentric pattern until it breaks free to orbit the Sun again. 2006 RH_{120} orbited the Earth four times in 2006–2007 while 2020 CD_3 may have arrived in 2016 and escaped Earth's gravity in spring 2020. Unlike the relatively stable martian moons (or asteroids), these temporary mini-moons are only a few meters (feet) across.

Earth captured another 'mini-moon,' (2020 SO) in late 2020, which may be gone by the time you read this. Differing from the natural ones, 2020 SO has a distinctly terrestrial origin. NASA confirmed that this satellite is a lost 1960s-era rocket booster from the Surveyor 2 lunar mission!

For the first pictures of an asteroid-belt body *in situ*, the NASA Galileo mission *en route* to Jupiter imaged two asteroids on its way: Gaspra[14] in 1991 and Ida[15] in 1993. Astronomers classify both bodies as S-type asteroids because they are primarily made of silicates. However, Gaspra appears to have more metallic components while Ida is mostly stony (Figures 9.15 and 9.16).

FIGURE 9.14   Tumbling asteroid Eros as imaged by the NEAR space probe.

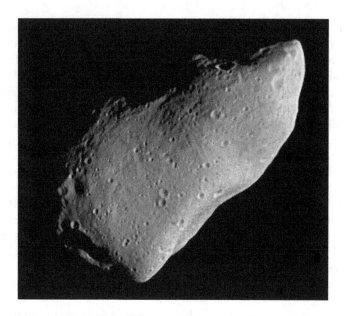

FIGURE 9.15 Gaspra imaged by the Galileo Jupiter orbiter.

NASA had already demonstrated that the Asteroid Belt poses little danger to space missions when Pioneer 10 and 11 passed safely through on their way to the outer Solar System. The asteroids are not clustered so closely that a spacecraft would have to swerve and dodge them continuously as in *Star Wars*.

Scrutinizing the images of Ida revealed that it has a moon, Dactyl.[16] Astronomers had previously thought such pairing possible but unlikely. Now, more than 250 such pairs are known plus a few asteroids with multiple moons. Binary asteroids exist as well. In a binary system, asteroids of similar mass orbit one another. In 2018, radar observations of 2017 YE$_5$ revealed that it was not a twin-lobed rock but rather two rocks, each about 0.9 kilometers (0.6 miles) across, circling their common center of mass roughly once a day. However, their colors suggest these rocks have different chemical compositions.

Yet, while seeing pictures produced a wealth of information about these bodies, visiting is always more interesting. NASA landed a probe, the Near Earth Asteroid Rendezvous-Shoemaker [NEAR Shoemaker[17]] mission, on 12 February 2001. The target asteroid Eros[18] is roughly cashew-shaped with a dent, or saddle, in its back. It measures only about 13 kilometers by 13 kilometers by

FIGURE 9.16 Ida imaged by the Galileo Jupiter orbiter. Notice its tiny satellite Dactyl to the right.

33 kilometers (8 miles by 8 miles by 21 miles); the need for three measurements emphasizes that it is definitely not round, like a ball or planet. It takes 5 hours, 16 minutes to rotate once around its axis. Its surface is dimpled with craters, strewn with boulders, and covered with regolith. Its chemical composition varies with location. Unraveling these clues may reveal its history and provide insight into the evolution of the Solar System.

Going one better, JAXA landed Hayabusa[19] twice on another stony asteroid Itokawa,[20] in November 2005, to collect samples. Although technical problems plagued the mission, it returned 1,500 minute particles of the dust that cover this asteroid to Earth on 13 June 2010, for study. Itokawa is a rubble pile loosely held together in a shape that kind of resembles a sea otter. Its surface has few craters; any impact is also likely to loosen material that shifts and flows down into the depression (Figure 9.17).

Analysis of the dust identifies S-type asteroids as the source of the meteorites most commonly found on the Earth. Some dust particles indicate that Itokawa must have once been part of larger body because they formed at a high temperature and cooled slowly. Individual dust grains also have microscopic craters from being hit by even tinier bodies. In addition, some particles contain water trapped within their crystalline structures opening the possibility that meteorite impacts contributed water to the terrestrial planets, which would have boiled away from the molten surfaces of planets in formation.

Building on the success of Hayabusa, JAXA launched its successor, Hayabusa 2, in 2014, which rendezvoused with the small near-Earth asteroid Ryugu[21] (C-type) in 2018. In addition to surveying Ryugu for more than a year and deploying three rovers, Hayabusa 2 completed its primary mission of collecting subsurface samples and returning them to the Earth on 6 December 2020. After the successful return of its 'treasure box' (sample capsule), JAXA extended Hayabusa 2's mission to visit other asteroids in the next decade.

NASA's Dawn mission orbited Vesta in 2011. The second largest body in the Asteroid Belt is very bright. Dawn imaged an impact crater so deep that it penetrated to a darker mantle. Thus, Vesta is differentiated. Dawn, then, moved on to an even more fascinating target (Figure 9.18).

Ceres, alone, accounts for 1/3 of the mass in the Asteroid Belt. Intriguingly, Dawn, orbiting the dwarf planet since March 2015,[22] discovered salts and mineral deposits on the surface that indicate that Ceres was once a water world. It now harbors a mantle of ice beneath a thin rocky crust. Ceres has relatively few craters. Astronomers speculate that ice volcanoes may have erased the results of

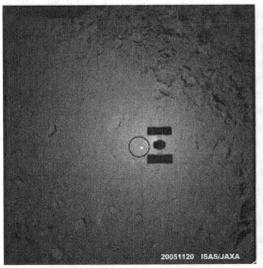

FIGURE 9.17   Hayabusa lands on Itokawa. Notice the spacecraft shadow on the surface.

FIGURE 9.18   Surface of Vesta as imaged by the Dawn space mission. Note the layering within the large crater.

earlier impacts or that movements in the low-density mantle smoothed them away. In addition, Dawn found ammonia on the surface indicating that Ceres may have formed farther away from the Sun than Jupiter and, then, migrated inward to its current location. Dawn also detected organic molecules on Ceres, which combined with water makes this dwarf planet a new target in the search for extraterrestrial life (Figure 9.19).

In addition to the Asteroid Belt, space rocks are also found sharing an orbit with some of the planets, most notably Jupiter. These asteroids occupy points approximately 60° ahead of and 60° behind the planet, as though they were opposing army camps locked in a stalemate; these are the L4 and L5 Lagrange Points. ➤ Chapter 2 ➤ Jupiter hosts thousands of these bodies. The occupants of its leading 'camp' are named for Greek warriors, while those of the following 'camp' are named for Trojan warriors, in a stand-off much longer than the fabled Trojan War.

So far, astronomers also have detected Trojan companions to Earth, Mars, Neptune, and Uranus. The one known Trojan asteroid leading the Earth in its orbit is 2010 TK7. Orbiting the Sun along with the Earth in this Lagrange point, it is no threat to us. With a diameter of about 300 meters (1,000 feet), it is smaller than the moons of Mars.

In summary, most asteroids stay far from the Earth in the Asteroid Belt, with some others occupying stable 'Trojan' orbits,[23] while most of the meteoroids that do hit the Earth are comparatively small. However, 'most' is not the same as 'all.'

Because a major impact could be devastating, in 1998, NASA set a goal of identifying 90% of the objects larger than 1 kilometer (0.6 mile) that come closer to the Sun than 1.3 au. Scientists estimate that such NEOs strike the Earth only about once in 500,000 years; although extremely rare, these events can cause a global catastrophe resulting in the death of a significant fraction of the world's population.

Having reached its first goal, NASA is now intent upon detecting 90% of NEOs larger than 140 meters (450 feet) in diameter. It estimates that about 25,000 of these exist. Scientists consider the smallest space boulders and short-period comets targeted by this search are capable of localized destruction, like the 1908 Tunguska Event mentioned earlier; such objects hit the Earth around once every few hundred years. As of 2020, the census of smaller NEOs was about 50% complete.

FIGURE 9.19 Ceres imaged by the Dawn space mission.

With thousands of potentially dangerous asteroids and other bodies not yet found, keeping our telescopic eyes peeled is worthwhile. But, what to do when we see one coming? The emerging field of planetary defense is studying our options for preventing a collision, if possible, or reducing the devastation, if not. To prevent a collision, we just need to adjust the orbital speed of the asteroid or comet around the Sun slightly. As discussed ➤ Chapter 3 ➤, a particular orbit has a particular average speed associated with it; with a few years lead time, changing that speed by as little as centimeters per second could alter the orbit enough to eliminate the threat.

Using one of those nuclear-tipped missiles—and this is the only good use of nuclear missiles we can imagine—is one possibility. However, instead of hitting the asteroid or comet directly, NASA would explode the device just above the surface, which would vaporize the surface on that side. The escaping material nudges the asteroid in the opposite direction. We have to be careful, though: Accidentally breaking an asteroid into pieces could result in an even worse 'shotgun blast' to the Earth! Two other less dramatic options would involve either actually hitting the asteroid with a space probe to prod it or using the space probe's weak gravity to tug it out of the way. NASA takes all this very seriously and is planning a mission to test such approaches. ☛ Chapter 17 ☛

In the meantime, the OSIRIS-REx mission began orbiting Bennu,[24] a potentially hazardous asteroid, in 2018. It comes close to the Earth every 6 years with a chance of hitting late in the 22nd century. It is the smallest solar-system body explored with a diameter of 510 meters (1,700 feet). However, if it struck land, Bennu could devastate the area within about 50 kilometers (30 miles), which is larger than the US state of Delaware, with detectable effects out to about 500 kilometers (300 miles), which is larger than Texas. OSIRIS-REx's mission is to collect a sample from this carbon-rich asteroid and return it to Earth in 2023.

In late 2019, NASA designated the Nightingale site as the primary OSIRIS-REX landing spot. In 2020, the orbiter gathered its sample: It was so successful at doing this that its collection chamber overflowed with asteroid material!

Understanding NEOs and preparing for potential collisions are vital to our self-interest as a species. As our experience with COVID-19 teaches us, the cost of studying uncommon but cataclysmic elements in nature is money well spent (Figure 9.20).

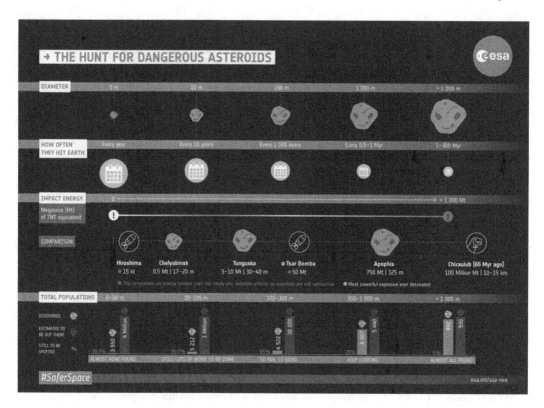

FIGURE 9.20 Asteroid risk assessment.

A future large impact could radically change life on the Earth as evidence suggests that one has already done so. As described earlier, the K-Pg extinction event, during which the dinosaurs disappeared 66 million years ago, was likely the result of the collision of the Earth and an asteroid (or comet). At the stratum below ground, corresponding to this date, a world-wide layer of rock is enriched with the chemical element iridium.[25] Iridium is rare in the Earth's crust but not in asteroids.

The scenario goes like this: 66 million years ago, an asteroid fell into the Earth, raising a cloud of dust all over the world. The heat of the explosion caused a global forest fire. Dust and soot blotted out sunlight, sudden climate changes ensued, and acid rain helped kill off most of the Earth's plant and animal life. The massive Chicxulub crater just offshore of Mexico matches the impact date very well, and it was produced by an object big enough to do the job (Figures 9.21 and 9.22).

As devastating as the Chicxulub impact must have been, it is not the only creditable theory about what eradicated the dinosaurs. Paleontologists, scientists who specialize in fossils, also consider the possibility that the changing climate in response to plate tectonics and increasing volcanic activity could have caused a slower extinction. Perhaps, the dinosaur population was already dwindling due to terrestrial causes and an extraterrestrial impact just hastened the job. Scientists have also raised the possibility that a smaller impact around 13,000 years ago may have challenged the mastodons and giant sloths of North America, but that was also a period of fluctuating climate after the last ice age.

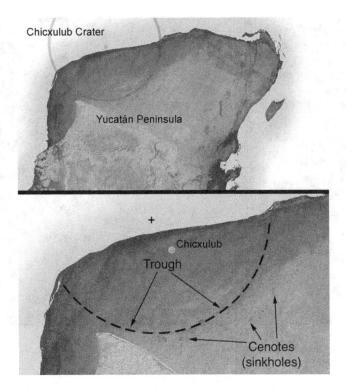

FIGURE 9.21   Location of the Chicxulub crater in the Yucatan Peninsula.

FIGURE 9.22   Forest fire, such as might be triggered by an asteroid impact.

Still if the impact had anything to do with the demise of the dinosaurs, we should thank our lucky star (asteroid): The absence of the dinosaurs left an ecological niche to be filled by the evolving mammals and eventually us.

But, of course, whatever impacts can provide, they can also take away. On 1 September 2017, Florence, a 4.4-kilometer (2.7-mile) diameter asteroid, came as near as 7 million kilometers (4 million miles) to the Earth. On 5 June 2020, a smaller asteroid (2020 LD), 122 meters (400 feet) in size, hurtled past us within 80% of the Earth-Moon distance. On 16 August 2020, astronomers watched asteroid 2020 VT4, perhaps as large as 11 meters (32 feet) across, zoom over the South Pacific a mere 400 kilometers (250 miles) high—the orbital altitude of the International Space

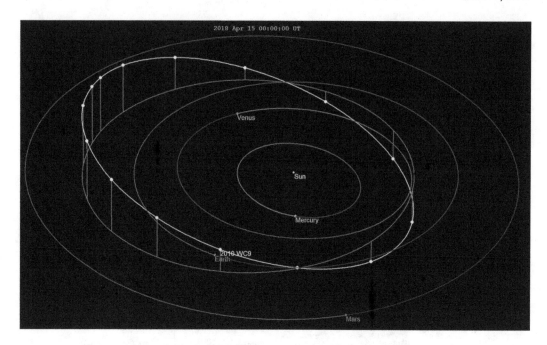

FIGURE 9.23   100-meter (300 foot) asteroid 2010 WC$_9$ passed the Earth at a distance closer than that of the Moon on 15 May 2018.

Station. And then there are those objects that astronomers entirely missed! Compared to the scale of the Solar System, these instances were close. Chicken Little was right! (Figure 9.23)

NOTES

1 Geologists abbreviate Cretaceous as 'K' for Kreide, which means chalk in German. Cretaceous comes from crēta, which is chalk in Latin.

2 The Minor Planet Center assigns provisional designations to asteroids and comets in order of discovery; in this case, this body was the 628th object discovered in the first half of October 2015. When enough observations are available to ensure its orbit is sufficiently well-understood, the Minor Planet Center will issue it a permanent number and allow its discoverer to name it, e.g., (7195) Danboice and (25153) Tomhockey. For ease of reading, we have disregarded the catalog numbers associated with named asteroids.

3 Named for Lewis Swift ⟨1820–1913⟩ and Horace Tuttle ⟨1839–1893⟩. who discovered it independently in 1862.

4 Named for Ernst Tempel ⟨1821–1869⟩ and Horace Tuttle, who discovered it independently in the mid-1860s.

5 Son of the Greek god Helios, who died while foolishly trying to drive the Sun chariot.

6 Interestingly, more documented cases exist of meteorites damaging cars than injuring people. To date, no authenticated report claims that a rock from space killed anyone.

7 Shoemaker is best known for his role in discovering Comet Shoemaker-Levy 9, which crashed into Jupiter. ☛ Chapter 13 ☛.

8 Sometimes called astroblemes from blêma, Greek for wound.

9 Also known as the Main Asteroid Belt.

10 Prototype named for an Egyptian Sun god, specifically the deified solar disk.

11 Prototype named for a Greek Sun god, sometimes distinct from Helios who drives the solar chariot.

12 Prototype named by Austrian astronomer Theodor von Oppolzer ⟨1841–1886⟩ in memory of his daughter.

13 Nurse to Zeus, who was the king of the Greek gods.

14 Crimean resort, for something completely different.

15 Nymph who cared for the Greek god Zeus as an infant.
16 Mythological creature supposed to live on Mount Ida in Crete.
17 NASA renamed this mission to honor Eugene Shoemaker, who died while the mission was en route.
18 Greek god of love.
19 Which means falcon in Japanese.
20 Named for Hideo Itokawa ⟨1912–1999⟩, the aerospace engineer who led Japan's early rocketry experiments.
21 A magical underwater Dragon Palace in Japanese folklore.
22 NASA officially ended the mission after they lost contact with the orbiter in late October 2018. However, Dawn should not crash into Ceres before 2038.
23 Other harmless asteroids include at least one that travels about the Sun entirely within the orbit of Venus.
24 Bird appearing in Egyptian creation myths.
25 Iridium is a transition metal with 77 electrons. The original concept for the Iridium communication satellite system used 77 active satellites, but it actually operates 66. Ever seen a bright flash of light in the sky? It may have been caused by an Iridium satellite.

10 Gas Giants: Jupiter and Saturn

The tracking of Jupiter and Saturn against the background stars began long ago. Our ancestors were patient enough to notice their motions, even though Jupiter requires nearly 12 years to orbit the Sun while Saturn takes more than 29 years. Cuneiform tables dating to the 4th century BCE record Babylonia observations of these planets, which were the most distant ones followed in ancient times. The Babylonians associated Jupiter with Marduk, the patron deity of their capitol city, who the Romans, in turn, identified with their chief god, Jupiter. The Babylonians associated Saturn with Ninurta, who was initially a Sumerian god of agriculture; the Romans could equate him with a mythical king who had brought agriculture and civilized behavior to his people (Figure 10.1).

In China, Gan De's ⟨circa 365 BCE⟩ observations of Jupiter also date to the 4th century BCE. Within 200 years, Chinese astronomers were familiar with the retrograde motion of five planets, including Saturn. In particular, they noted the motion of Jupiter through approximately one constellation of their zodiac each year[1]; consequently, they called it the 'Year-Star.' When combined with the five traditional Chinese elements,[2] the sequence creates a 60-year cycle that appears in imperial records starting in the 1st century. Another Chinese scholar, Liu Hsin ⟨circa 12⟩ proposed a rule adjusting the cycle every 144 years to account for the fact that Jupiter's actual orbital period is less than 12 years, but it was not widely adopted. Starting at Chinese New Year, 2021 is represented by a metal ox, followed by the metal tiger associated with 2022; after 12 years, the cycle continues with the water ox and water tiger of 2033 and 2034, respectively.

Ancient astronomers of any nationality charted the positions of Jupiter and Saturn and recorded their apparent brightnesses of the planets; these were the distinguishing characteristics of the known planets. However, telescopic observations beginning with Galileo and space missions beginning with Pioneer 10 have revealed how completely different these worlds are from our terrestrial experience.

The inhabitants of the outer Planetary System are the Jovian Planets after their archetype, Jupiter. More specifically, the Gas Giants are closest to the Sun; Jupiter is inarguably giant. At a distance of 5.2 au, Jupiter is such a major jump away from the Sun that sunlight takes 43 minutes, 10 seconds to arrive there (Figure 10.2).

The king of the planets has the largest entourage of smaller bodies orbiting it as befits his station. The present count, stands at 79,[3] but four large ones dominate because of their size and mass. More will be told about Jupiter's satellites ☛Chapter 12 ☛.

Jupiter is the most voluminous planet; 1,300 Earths could fit inside it. It holds more than twice as much mass as the rest of the Planetary System combined (equal to 318 Earths). It has been said that an objective inventory of the Solar System would be the Sun, Jupiter, and 'debris'! Jupiter would seem to dominate just about everything in our system of planets, except for one way in which it is a 'lightweight': The average density of Jupiter is about 1.3 g/cm^3. Recall that all the Jovian Planets have densities around the density of liquid water (1.0 g/cm^3), but assuming that their composition is mostly water would be a mistake! (Figure 10.3)

Obviously, Jupiter must be made of very low-density elements, of which the list quickly narrows down to two: hydrogen and helium. Specifically, 88% of Jupiter is made up of hydrogen and another 8% is helium. The remainder consists of all the other elements combined. The recipe for the remaining Jovian Planets is similar. The chemical make-up of Jupiter is also similar to that of the Sun.

These first two elements are not solid on the Earth; we think of them as gases. Most of our hydrogen is bound in the water molecules. Our remaining helium is physically trapped beneath the surface. While the Earth may once have had extensive atmospheres of hydrogen and helium, too, these gases have drifted away; the small gravity of Earth was unable to hold onto them.

FIGURE 10.1 Jupiter with over-exposed Moon.

FIGURE 10.2 Jupiter is always fully sunlit as seen from the Earth; this Galileo orbiter image shows it gibbous.

FIGURE 10.3   Jupiter and the Earth compared in size.

Jupiter can retain these elements because of its enormous gravity and the fact that the gases are not 'trying' very hard to escape. Jupiter has an effective temperature at the cloud tops of −140°C (−230°F). The molecules are moving sluggishly and do not have enough energy to escape. Higher up, where the temperature is greater, there are a few fast-moving molecules; Jupiter's strong gravity still prevents them from escaping.

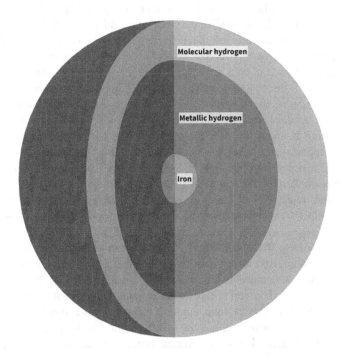

FIGURE 10.4   Jupiter in cross section.

Jupiter is another sort of planet and deserves another cross-section diagram. It begins with a thin, cold outer layer of gas, but as we go deeper (in our imagination) into the planet the atmosphere warms up and grows denser. It is the same here on the Earth. Traveling into the oceans of the Earth, we feel the pressure increase as we swim deeper. It makes our ears hurt. The same is true of Jupiter's atmosphere, the deeper, the more pressure (Figure 10.4).

As for the temperature, heat always travels from where it is warmest to where it is coldest. We can extrapolate that the center of Jupiter is very hot. Jupiter formed hot and still retains much of the heat with which it was born. With its surface area small compared to its volume, Jupiter makes a good insulating bottle. In fact, Jupiter radiates some heat. If we claim that the difference between a star and a planet is that stars radiate and planets do not, this is one more way in which Jupiter resembles the Sun.

Returning to our descent into Jupiter, the molecular hydrogen and helium get thicker and warmer. The temperature becomes 'comfortable' at a little less than 27°C (86°F), but only at a point where the atmospheric pressure is several times that at sea level on the Earth. Eventually, starting about 1,000 kilometers (600 miles) down, the helium-rich, molecular hydrogen gradually becomes a liquid.

The pressure increases even more such that, at a radius 80–90% of the planet and a pressure three *million* times that at the Earth's surface, Jupiter is squeezed so tightly that electricity can easily flow through the hydrogen. It becomes a very good conductor. Because it has the properties associated with metals, chemists call this strange kind of hydrogen: metallic hydrogen. Nobody has even seen an ocean of metallic hydrogen. However, laboratory experiments show that hydrogen must become metallic under these extreme conditions.

Finally, we reach the core of Jupiter consisting of rock and metal. It is small compared to the size of the planet but could be larger than the Earth with more than 10 times its mass. The jovian core boundary is 'fuzzy,' less distinct than that of a terrestrial planet. The core temperature of Jupiter soars to about 24,000°C (63,000°F), many times hotter than the visible surface of the Sun. The internal pressure exceeds 100 million times that of our own atmosphere. At these temperatures and pressures, molecules cease to be and dissociate into their elemental constituents. The element carbon may be crushed to its diamond form. The solution at the center of Jupiter is possibly the largest diamond deposit in the Solar System! Ah, but how to extract that wealthy resource? (Figure 10.5)

Planetary scientists infer the radii of the different layers of Jupiter (and other planets) by measuring their moments of inertia. As we discussed earlier, the same push on a skateboard produces greater acceleration than on a car because a car has more mass. Another way of thinking about mass is to consider it the resistance of an object to changing its motion along a line, or linear inertia. For a rotating object, its resistance to change in its rotational motion is its moment of inertia. If you remove a wheel from that skateboard, you can spin it easily around the mounting axis. However, flipping it about the disk, or perpendicular to the mounting axis, is awkward. Moment of inertia depends upon how mass is distributed through a rotating body. Have you ever balanced and twirled a pencil on your finger? A dinner knife? A spoon? The uneven shape of a spoon makes it more challenging; its moment of inertia is greater.

Observing how the trajectories of passing space probes deviate after close encounters with a planet reveals the planet's moment of inertia. This is where NASA's Juno[4] mission excelled. Through repeated close approaches as it orbits Jupiter, the planet's interior has been revealed in great detail. Initial results indicate that the internal structure is more complex—less divided into the distinct regions described above. Further analysis of the Juno data should clarify Jupiter's interior.

The Juno mission returned detailed and unprecedented images of Jupiter. Making highly elliptical orbits, it passes within several thousand kilometers of its poles each time it makes a closest approach to Jupiter, about every 50 days. Before Juno, Jupiter's polar regions had only been glimpsed. It is a chaotic place of clusters of storms without the characteristic bands typical of Jupiter. Cyclones collect into polygonal structures known as 'vortex crystals' at Jupiter's north and south poles. Lightning is often concentrated there and the source of very energetic electron emissions.

Comparing pressure

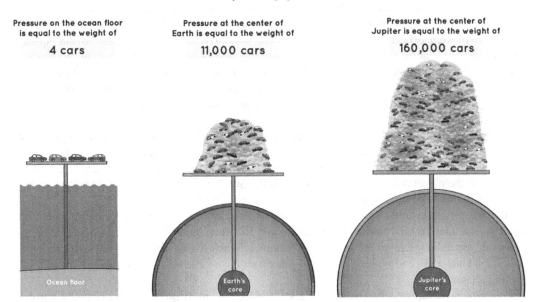

Pressure on the ocean floor is equal to the weight of	Pressure at the center of Earth is equal to the weight of	Pressure at the center of Jupiter is equal to the weight of
4 cars	**11,000 cars**	**160,000 cars**

FIGURE 10.5   Pressure inside of Jupiter.

Juno also found that the low-latitude banded structure penetrates deep within the atmosphere, at least 3,000 kilometers. It measured the amount of water in Jupiter's atmosphere to be almost three times that of the Sun, much greater than the value reported by the Galileo probe in 1995. However, the atmosphere is not well mixed even well below the cloud tops, a puzzle to astronomers. In addition, Jupiter's gravity and magnetic fields are lopsided.

Have you noticed something peculiar about Jupiter yet? If you were intent on becoming the astronaut who first plants your flag on Jupiter, you would descend through its atmosphere, into its liquid layer, through its metallic hydrogen layer, and on to its (presumably molten) core. There is no place to stand on Jupiter, no place to stick the flag! Jupiter is just a big ball of fluid, with no solid crust—much different from the terrestrial planets. It deserves its nickname, a 'Gas Giant.' However, if you stood on a platform at the top of the atmosphere, you would feel very heavy, 2.5 times your weight on the Earth, due to Jupiter's strong gravity. It is not a place to visit if you desire weight loss!

Why does a Gas Giant neither expand nor contract, like a balloon? Imagine Jupiter as an onion of many thin layers. The top layer has nothing to do. However, the next layer does have a job: It must hold up the layer above, which gravity is trying to pull downward. The second layer supports the first layer by having a higher pressure. (Think of an automobile, which may be raised above the ground simply by pumping up the pressure in the tires high enough.) The third layer of Jupiter must support the first and second layers. Its pressure must be higher than that of either layer atop it in order to accomplish this. If you have ever been at the bottom of a pile-on, you know how that feels!

On and on into the planet, as long as the material involved is a compressible fluid, the pressure in Jupiter must increase, in a specific way, in order to keep each layer in balance. If this were not the case, Jupiter (and other Gas Giants) would either expand (the pressure with depth is too great) or contract (the pressure with depth is too low). The balance between pressure and weight is called hydrostatic equilibrium.

The Jovian Planets may be *slightly* out of hydrostatic equilibrium. If a planet should shrink, even a little, the falling layers would generate extra heat. This heat combines with the residual primordial heat of the planet to make it still hotter as a function of depth. Nevertheless, at most, Jupiter contracts a mere 2 centimeters per year (1 inch per year).

Now, with Jupiter so massive and sizeable, you might assume that it has the right to spin slowly on its axis, that is, until you consider the large amount of matter that has contracted to form the planet—pulling its 'arms' in like the skater. Jupiter rotates in approximately 9 hours, 55 minutes, the fastest rotation period among the planets. (The measured rotation rate at the jovian equator differs from that near the poles by 6 minutes: While this sounds peculiar, remember that Jupiter is a fluid world.) It is this rapid rotation that helps whip up Jupiter's magnetic field at the core/metallic-hydrogen boundary. This field is the strongest of any planet in the Solar System—40,000 times stronger than the Earth's. It is intense enough to produce radio noise that astronomers can pick up on the Earth (Figure 10.6).

The region enveloping a planet, under the influence of that planet's magnetic field as opposed to that of the Sun, is called its magnetosphere. Jupiter's magnetosphere is truly huge, 20,000 times greater in volume than that of the Earth's. It extends nearly 3 million kilometers (about 2 million miles) from the planet and boasts a 'tail' opposite the Sun 1 billion kilometers (600 million miles) long. It is the largest permanent feature in the Planetary System.

Jupiter's magnetic field traps electrically charged particles from the Sun and some sputtered off Jupiter's satellites in tori[5] surrounding Jupiter just like the Van Allen Belts encircle the Earth. However, the radiation level in the jovian versions is much higher—a thousand times the lethal dose for a human! Even radiation-hardened spacecraft have suffered damage passing through the jovian radiation belts.

Less violently, Jupiter's magnetic field also produces the most powerful and brightest aurorae on any planet, seen in Jupiter's polar regions. These manifestations are tens of times more energetic than those beautiful, twisting undulations that appear from time-to-time on the Earth. However, how they work is still unsolved.

Unlike the Earth, the jovian fireworks never cease and have an additional source: particles escaped from fiery volcanoes on the planet's satellite Io. ☛ Chapter 12 ☛ While the gases in the Earth's atmosphere glow red, green, and purple, the color of Jupiter's light show is mostly ultraviolet due to its hydrogen content. Thus, unfortunately, a tourist near Jupiter could not witness it without special goggles (Figure 10.7).

FIGURE 10.6 Jupiter's magnetic field illustrated. The field is very intense even beyond the orbits of Jupiter's inner satellites.

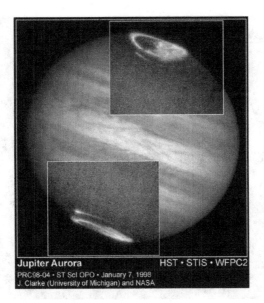

Jupiter Aurora   HST • STIS • WFPC2
PRC98-04 • ST ScI OPO • January 7, 1998
J. Clarke (University of Michigan) and NASA

FIGURE 10.7   False-color image of Jupiter's aurorae.

Let us return to the outmost levels of Jupiter that we can see. Hydrogen and helium are transparent. What we are viewing through a telescope are clouds made of ammonia and other available molecules. Water clouds may lurk slightly beneath.

Images of the visible part of the atmosphere demonstrate yet again Jupiter's fluid nature: The planet is slightly out of round! The diameter measured through the equator is lengthier than that measured through the poles: 143,000 kilometers (89,000 miles) versus 134,000 kilometers (83,000 miles). Jupiter's rapid rotation 'sloshes' Jupiter outward at the equator turning it into an oblate spheroid. This extra force that increases towards the equator and pushes out a bulge is called centrifugal force, arising from the planet's rotational inertia. It is exactly what you feel as you lean into a curve while driving or hold on to a merry-go-round to keep from being thrown off.

When we write of Jupiter's visible atmosphere, we are not referring to floating clouds over a surface. Instead, we mean cloud layers covering cloud layers covering still deeper cloud layers. For instance, the zebra-like appearance that Jupiter most often displays consists of white clouds covering, at certain latitudes, deeper, browner layers of more clouds below (Figure 10.8).

Jupiter's rotation, and the strong upper-atmospheric winds that accompany it, stretch the visible cloud layers into ribbons parallel to the equator. The winds blow at fantastic speeds more than 540 kilometers per hour (340 miles per hour), faster than any gale on the Earth. Rather arbitrarily, the white ones are named 'zones' and the dark ones 'belts.' The pattern is wider near the equator than it is near the poles. From the Earth, the belts and zones always appear straight—like the stripes on a rugby shirt—because Jupiter has an obliquity of only 3.1°. Occasionally, a band can switch from zone to belt or *vice versa*, demonstrating the superficial nature of these features. This back-and-forth albedo ornamentation only breaks down into a jumble of twisted, colored clouds near the poles (Figure 10.9).

Observations show that the zones represent updrafts of material from layers below, while the belts are downdrafts. In these bands, convection moves heat from Jupiter's interior to its surface. Large parcels of hot, less-dense gas rise from below and release their heat. Cooler, more-dense parcels sink to be heated as they descend. This recycling of material is very similar to processes in the Earth's atmosphere, such as thunderstorms (Figure 10.10).

The banded appearance runs deep in Jupiter's atmosphere; it may represent the tops of nested cylinders that go all the way to metallic hydrogen layer. Ever-changing weather at the top is driven mainly from within. Internal energy fuels these weather systems rather than external sunlight like

FIGURE 10.8 We are used to seeing Jupiter in the plane of the ecliptic. This Juno orbiter image from 'beneath'; presents a view never seen from the Earth.

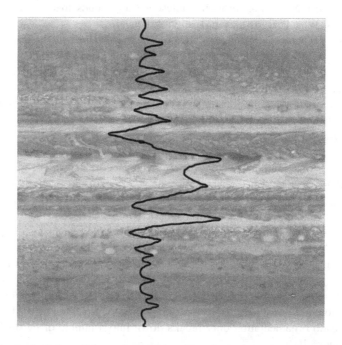

FIGURE 10.9 Graph of relative wind speeds on Jupiter. Note the horizontal peaks in the black line represent strong winds, which change direction at the boundaries between belts and zones.

on the Earth; sunlight is dim at Jupiter's distance. However, studying the properties of Jupiter's dynamic atmosphere gives us insight into our own turbulent atmosphere.

When writing of terrestrial planets, we borrowed some words from geology. Now with Jupiter, we appropriate terms from meteorology. The zones correspond to high-pressure areas on the Earth, the belts low-pressure areas. On Jupiter, these patterns have been stretched around the entire

FIGURE 10.10 Juno image of the border between a zone and belt.

planet. The belts and zones are bordered by jet streams blowing east and west. Traces of ammonia will rise in the zones to the top of the troposphere, freezing into an opaque layer of ice crystals about 50 kilometers (31 miles) thick. However, it is pushed downward in the belts, through a clearing layer, and the atmosphere remains transparent, allowing us to see a brown layer of cloud made of unknown aerosols. Occasional holes permit us to spy an even deeper, warmer cloud layer that appears blue and glows in infrared light.

The Galileo Probe parachuted into Jupiter on a one-way mission during which it provided information about the third dimension unavailable in photographs. The robotic measurements largely correspond with the model described above (Figure 10.11).

Within the white-and-brown cincture of Jupiter are a multitude of smaller features. White ovals appear and disappear, presumably upwelling to an altitude at which ammonia freezes. Elsewhere, brown spots, unfortunately named 'barges,' mark places where the upper-level clouds have been pulled away, revealing the underlying brown level (Figure 10.12).

While many ovals come and go, and the same is true for the barges, Jupiter's visible atmosphere has one long-lasting and unique feature: a great red spot named unimaginatively the 'Great Red Spot'. There is only one of these, and it is certainly great. It has been 14,000 kilometers (8,700 miles) wide and thrice that long—big enough to swallow two Earths. Astronomers have observed the Great Red Spot for almost four centuries, dimming and brightening, shrinking and growing. Discovered in 1665 by Cassini, it is sometimes hard to see, sometimes impossible to miss. Yet, it is always there, rolling between the wind jets that also constrain the belts and zones. Lately, it has been getting smaller, but taller. Nobody knows why (Figure 10.13).

Few other jovian spots have the Great Red Spot's particular pink tinge. The agent responsible for its color may be a dredged-up layer of sulfur compounds, or organic molecules like those that give apples or tomatoes their hue; all of which can appear red. The color may also be due to the record-breaking height that the Great Red Spot achieves, 8 kilometers (5 miles) above the ammonia cloud tops. Here, poorly understood photo-chemical reactions occur, resulting in a soup of complex molecules.

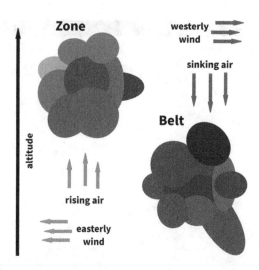

FIGURE 10.11   Diagram comparing a zone and belt in the third dimension.

FIGURE 10.12   Image of the jovian south pole made by the Juno orbiter.

This and other rolling features on Jupiter are nothing less than vast, fierce storms in the jovian atmosphere. Astronomers debated the nature of the Great Red Spot for centuries but have now determined that it is the 'mother of all cyclones,' rotating counterclockwise in 6 days.

Completing this stormy image, space probes have photographed tremendous flashes on Jupiter's night side, lightning bolts with the energy of a nuclear bomb! Sprites and elves are found at higher altitudes. Not the capricious creatures of English folklore, these manifestations are brief but intense electrical outbursts associated with the lightning below. Humans have imaged sprites and elves on the Earth for three decades but have only recently seen them on other planets (Figure 10.14).

In summary, forecasting the weather on Jupiter is easy: storms today, storms tomorrow, storms a hundred years from now.

Far above the clouds and storms of Jupiter circle a faint, insubstantial ring system of interplanetary dust in Jupiter's equatorial plane. With a legion of forces acting upon them, the ring particles probably should disperse in only a few thousand years. Therefore, they must be

FIGURE 10.13   Great Red Spot imaged by the Juno orbiter.

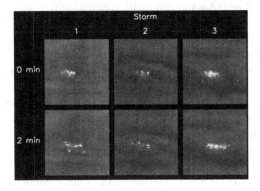

FIGURE 10.14   Lightning storms on Jupiter.

replenished if the ring system is a permanent feature. Interplanetary meteoroids, attracted by Jupiter's powerful gravity, pound into the small inner satellites of Jupiter. They chip off tiny fragments that go into orbit around the planet, forming the rings. Ripples in the rings may be sites where meteoroids passed through (Figure 10.15).

The main ring is only 30–300 kilometers (19–190 miles) thick, but it is 6,500 kilometers (4,000 miles) wide. From the middle of it emerges a cloud of material called the halo ring that expands in thickness down to the jovian cloud tops.

The three well-named gossamer rings are made of minute flecks from Jupiter's small satellites, Amalthea, Thebe, and Adrastea; each particle is the size of those that make up cigarette smoke. These thin rings are thicker and reside farther out, extending 130,000 to 226,000 kilometers (80,000 to 140,000 miles) from the center of the planet.

Since Galileo first spied Jupiter, astronomers have wondered what it would look like up close. NASA's plan for outer planetary exploration began with space-probe flybys, followed by orbiters. Nine spacecrafts have visited the jovian system. The twins, Pioneer 10 and 11, passed by in 1973 and 1974. They were followed by the more-sophisticated Voyager 1 and 2 in 1979. NASA's orbiter Galileo arrived in 1995 and ended its mission in 2003. These missions returned a wealth of

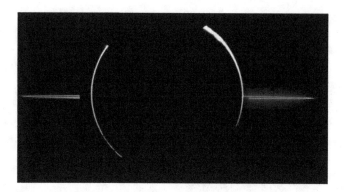

FIGURE 10.15 In this Juno image, Jupiter's rings show up best backlit, much like the dust on an automobile wind shield.

information and vastly improved our knowledge of Jupiter, but raised so many new questions that NASA decided to send a second orbiter, Juno, arriving in 2016 and currently in orbit. It has yielded spectacular results.

Where are the earlier space probes now? Utilizing the mass of Jupiter as a gravitational 'slingshot,' the Pioneers and Voyagers achieved escape speed from the Solar System, so they are headed for interstellar space, along with the New Horizons mission that glanced at Jupiter on its way to Pluto. ☛ Chapter 14 ☛ NASA intentionally crashed Galileo into the giant planet in 2003 to avoid possible biological contamination of the jovian satellite system. As required by planetary protection guidelines, NASA plans to de-orbit Juno into Jupiter's atmosphere, again minimizing the risk of contamination and avoiding space debris.

These missions, complemented by ground- and space-based observations, such as those made by HST, have increased greatly our understanding of the Jupiter system in a mere human generation.

This brings us to Saturn, the farthest planet from the Sun that was spied by the ancients. To the Mesopotamians, it was 'The Oldest of the Old.' The Hindi incarnation Shani judged peoples' good and bad deeds. The Greek Cronus, he was the father of Zeus.[6] The modern name is the ancient Roman deity who ruled the world during a mythic period of prosperity. Gift-giving and partying characterized the winter festival in his honor, the Saturnalia.

At 9.5 au, Saturn is nearly twice as far from the Sun as Jupiter. Sunlight takes 1 hour, 19 minutes to travel this far. Its greater distance from the Sun produces the major differences between the two planets.

Saturn is the other Gas Giant. It is the second-largest planet; 760 Earths would fit inside it. Like its bigger neighbor Jupiter, it has an abundance of satellites[7] ☛ Chapter 12 ☛, a magnetosphere, and a ring system. Its atmosphere is composed of molecular hydrogen gas, which surrounds a larger liquid body of molecular and metallic hydrogen. At its core is probably a high-pressure, molten mass of rock and metals at a temperature of about 12,000°C (21,000°F). Although this structure resembles that of Jupiter, Saturn's layers of molecular hydrogen gas, liquid molecular hydrogen, metallic hydrogen, and so forth must be of different relative radii (Figure 10.16).

Saturn is less dense than Jupiter; in fact, it has the lowest density of any planet, eight times less than the Earth. So, yes, if you could find a hot tub big enough, you could float Saturn in it. (And upon removing, it would leave a ring!)

With such a low density, Saturn's chemical composition should be similar to Jupiter's. Based on the upper atmosphere, Saturn *appears* to be depleted in helium with 96% hydrogen and 3% helium. This region of Saturn may be so cold that helium droplets have formed and precipitated into hidden layers below. Imagine helium rain! Like Jupiter, Saturn radiates more energy than it receives from the Sun. Falling helium may provide some of this excess energy. Just as water falling into a canyon heats up, so does the falling helium. One helium atom is not very massive; however, it has lots of

Jupiter & Saturn
Subaru Telescope, National Astronomical Observatory of Japan

CAC (B, V, & R)
January 28, 1999

FIGURE 10.16   Jupiter and Saturn compared through a ground-based telescope.

friends. This rain of helium droplets falling on and saturating the liquid hydrogen ocean below could be a significant contributor to Saturn's internal heat.

Receiving only 1% of the sunlight that we receive at Earth, the temperature at the top of Saturn's cloud layer remains well below zero centigrade, about −180°C (−288°F). This is so even though its atmosphere also is heated from below like Jupiter. Astronomers assume saturnian clouds are made of the same ingredients as jovian clouds. However, images of Saturn lack the colorful clouds seen on Jupiter: In this more intense cold, a layer of uniform, lightly tinted haze forms and obstructs our view.

Without the use of spacecraft and the most-modern, Earth-based telescopes, very few distinct cloud features on Saturn are seen at all. Just one easily observable stormy outbreak forms regularly. It does so every saturnian year when upwelling brighter material punches through the haze layer. This so-called 'Great White Spot' occurs at the same northern latitude each appearance. Astronomers look for it in 2020–2021 at Saturn's northern summer solstice (Figure 10.17).

Saturn, too, is a fast rotator, almost 11 hours per saturnian day on average. This is only 44 minutes longer than Jupiter. With a mass—and 'surface' gravity—smaller than Jupiter's, Saturn's equatorial bulge is greater with an equatorial diameter of 120,000 kilometers (75,000 miles) versus a polar diameter of 110,000 kilometers (68,000 miles). Like Jupiter, Saturn's visible atmosphere spins differentially, faster at the equator and slower as one moves to the poles. Its rapid rotation also draws its cloud patterns into white and brown bands, wider near the equator and narrower near the poles. However, the saturnian ones are far less distinct than those on Jupiter. Atypically, Saturn comes out ahead of Jupiter in one parameter: Its ferocious upper-atmospheric winds blow more swiftly. They reach speeds of 1,800 kilometers per hour (1,100 miles per hour), which is four times faster than any recorded on Jupiter, or ten times faster than an Earthly typhoon! (Figure 10.18)

During observations from 2004 to 2017, NASA's Cassini orbiter imaged a hexagonal arrangement of clouds rotating at Saturn's north pole in high resolution. It appears to be a wavy, circumpolar wind jet moving at 320 kilometers per hour (200 miles per hour) around a vortex. The whole structure could fit four Earths within its boundaries. A so-called 'standing wave,' there is nothing else like it in the Solar System (Figure 10.19).

FIGURE 10.17 Last apparition of the Great White Spot as seen in a Cassini orbiter image.

FIGURE 10.18 Oblateness of Saturn.

Meanwhile, at Saturn's south pole, Cassini imaged a seemingly permanent hurricane, spinning at 500 kilometers per hour (310 miles per hour). It raises the local temperature to −120°C (−180°F). This is the only other place than the Earth that exhibits an eyewall in such a storm.

Saturn's magnetic field is 'only' 600 times as strong as the Earth's. It is almost perfectly aligned with the rotation axis. Nonetheless, Saturn's rings and many of its plenitude of satellites lie within its magnetosphere. The planet's radiation torus is fed, not so much by the distant Sun, but by particles from its rings and satellites. Like Jupiter, it also supports aurorae and produces radio emissions.

Saturn's defining feature is its beautiful set of wide, white, bright rings. Together they are as wide as 4½ Earths. Because Jupiter and the other jovian planets have ring systems that girdle

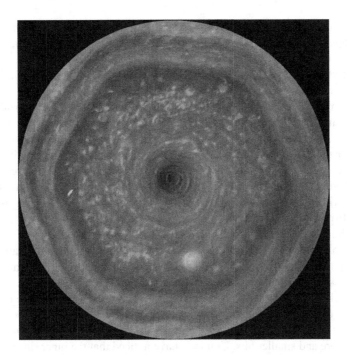

FIGURE 10.19 Hexagon storm at Saturn's north pole. This Cassini orbiter image has been color coded to emphasize the feature's shape.

their equators, Saturn is not unique. While each jovian ring system is special, all pale in comparison to Saturn's.

Galileo could not resolve Saturn's unusual form through his telescope in 1610. The planet looked like it had handles or arms, one on either side. He said it has 'ears' and wondered if they were a pair of odd satellites. To confound the mystery, sometimes the rings could be seen, sometimes not (Figure 10.20).

Armed with a better telescope, Huygens discerned that at least one ring encircled Saturn, not touching it at any point. Why, then, do Saturn's rings play Peek-a-Boo? Saturn has an obliquity of

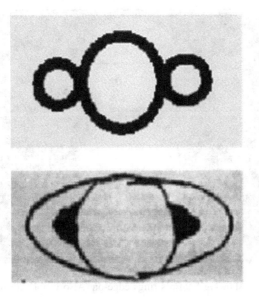

FIGURE 10.20 Galileo Galilei's drawings of Saturn (1610 [top] and 1616 [bottom]).

27° so twice a saturnian year, one of Saturn's poles is tipped toward us. At these times, astronomers see the rings at their greatest apparent width. However, a quarter saturnian year later, they see the rings edge-on. Rather, they do *not* see them—they are too thin to be resolved from the Earth. Hence, Saturn does 'now you see the rings, now you don't,' every 15 Earth years.[8] They will be at their apparent narrowest again in 2025 (Figure 10.21).

Saturn's rings are 282,000 kilometers (175,000 miles) wide. Saturn and its rings together take up a little less than the distance between the Earth and Moon. However, the rings are delicate, perhaps only tens of meters in thickness in the inner region and flaring out to thousands of kilometers at their outer edge but always densest in their mid-plane. To scale, it is as if someone had spread a sheet of onion-skin paper over many, many football fields (Figure 10.22).

Late in its mission, the Cassini orbiter flew between the Saturn's upper atmosphere and the inner edge of its rings. It found exceptions to the otherwise, largely flat ring plane. In an undetermined number of places, the disk is interrupted by mysterious 'massifs,' piles of material more than 1.5 kilometers (1 mile) high (Figure 10.23).

The rings are not continuous, rather they are made of billions of individual particles, ranging in size from that of dust, to sand, to pebbles, to boulders, to buildings. A few are mountain-size. You might want to think of ring particles as tiny 'moonlets.'

Each ring particle obediently obeys Kepler's Laws. They move fastest at the rings' inner edge, taking only about 6 hours to orbit the planet, and move progressively slower further from the planet. At the outer edge, particles complete an orbit in about 14 hours. Throughout, no doubt they collide, creating more and smaller bits. You may have noticed that the inner ring particles orbit at a speed faster than the planet rotates!

How do we know that the rings consist of particles? When the Voyager 2 space probe flew by Saturn on its way to Uranus, its radio signal to the Earth passed through Saturn's rings. The scatter of this signal was consistent with the ring particle theory.

Ring or rings? There is one gap easily visible to even modest Earth-based telescopes of today. Named the Cassini Division, after the astronomer who first spotted it, the interstice is 4,800 kilometers (3,000 miles) wide. So, there must be at least two rings. In fact, astronomers have known for centuries that there were at least three rings detached from the planet, named A, B, and C in the order in which they were discovered.

More rings have been found, so currently the order from planet outward is: D, C, B, Cassini Division, A, F, G, and E, with an insubstantial ring co-orbiting Saturn with its satellite Phoebe

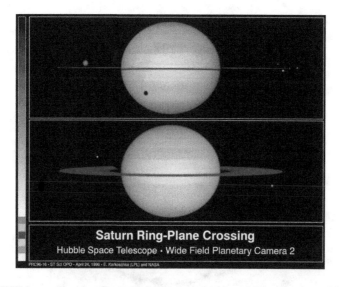

FIGURE 10.21   HST image showing Saturn's rings edge-on. Also visible are some of Saturn's satellites.

FIGURE 10.22   Rings of Saturn as imaged by the Cassini orbiter.

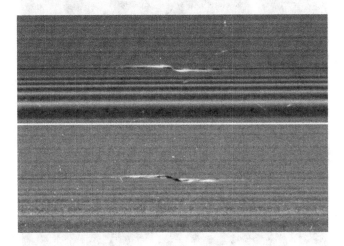

FIGURE 10.23   'Pile' in Saturn's rings imaged by the Cassini orbiter.

much farther out. However, in truth, robotic missions visiting Saturn show clearly that there are *thousands* of rings, sometimes called ringlets. They exhibit amazing complexity and beauty sculpted by gravity. The named rings are really groups of ringlets.

Small satellites of Saturn orbiting *within* the rings maintain some of the ring gaps. The gravity of a pair of so-called shepherd satellites conspires to keep the ring particles in their own ring (Figure 10.24).

Close inspection of the F Ring reveals that it appears braided! Kinks and bends in its ringlets give this illusion. The explanation is unknown (Figure 10.25).

Even though the rings look opaque, there is lots of space between the ring particles. Astronomers estimate that the total mass of Saturn's rings is less than a millionth of our single Moon.

The brightness of the rings is apparently due to their composition: ice (or rock and ice) covered to various degrees with dust. Just as the asteroids are examples of small terrestrial bodies,

FIGURE 10.24   Shepherd satellite 'patrols' the F Ring in this Voyager 2 image.

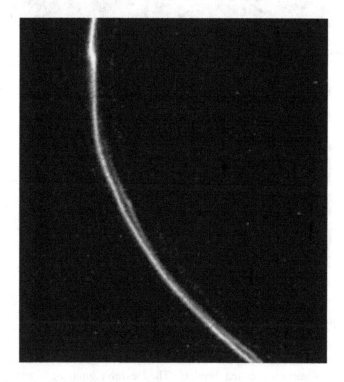

FIGURE 10.25   Saturn's 'braided' F Ring as imaged by Voyager 1.

the ring particles are examples of tiny *icy* bodies. We will describe not-so-tiny icy bodies later. ☞ Chapter 13 ☞

As viewed from one of the poles, dark 'spokes' sometimes radiate outward from the direction of the planet. They last for mere hours. If part of the ring system, these spokes should not maintain themselves for even that long; they should be torn apart by the different speeds of the rings. The spokes actually lie above the rings, consisting of dust propelled around the planet by Saturn's magnetic field. Is their source meteoroid impacts? Saturnian lightning? (Figure 10.26)

FIGURE 10.26   'Spokes' in Saturn's rings, as shown in a Voyager 2 image (left) and a Cassini image (right).

Where did the rings come from? Perhaps a prospective satellite, or a large comet or asteroid ventured too close to Saturn. If so, it would be pulled asunder by saturnian tidal forces, the difference between the stronger gravitational pull on its near side and weaker pull on its far side. Close to the planet, the tidal force can exceed the internal forces holding the prospective satellite together, resulting in destruction. French astronomer Edouard Roche ⟨1820–1883⟩ determined the farthest distance from the planet where this occurs is about 2 1/2 times a planet's radius. Within this Roche Limit, no satellite can coalesce gravitationally. Saturn's rings fit nicely within this region.

FIGURE 10.27   Final image of Saturn (backlit) by the Cassini orbiter.

Planetary rings may be a temporary phenomenon. In the case of Jupiter, the age of a ring particle is 100 to 1,000 years, so they must be continually replenished to last long. Saturn's magnetic field is dragging its ring particles into the planet. This Ring Rain will deplete the current set in 100 million years. What luck it is that we live at a time when we can see them through a telescope and enjoy them (Figure 10.27).

Four NASA spacecraft have visited the saturnian system: beginning with the flybys of Pioneer 11 in 1979 followed by the twins, Voyagers 1 and 2 in 1980 and 1981. Most recently, Cassini completed 13 years of observations. Similar to Jupiter missions, its sponsors intentionally crashed the orbiter into Saturn in 2017 when it was low on fuel; an action intended to prevent contamination. One probe is planned to return to the saturnian system. NASA's Dragonfly will launch in 2026 and explore Saturn's moon Titan with a rotorcraft in 2036. ☛ Chapter 17 ☛ .

Only one lonely space probe has visited the two Giant Planets we will now encounter.

NOTES

1 Rat, ox, tiger, rabbit, dragon, snake, horse, sheep, monkey, rooster, dog, and pig.
2 Metal, water, wood, fire, and earth.
3 Of which 78 are confirmed and 1 is a provisional (unconfirmed) satellite.
4 Named for Jupiter's queen.
5 Doughnut-shaped figures, the singular of tori is torus.
6 Cronian (or kronian) is also used to describe something related to the planet Saturn.
7 Present count is 53 confirmed satellites, with another 29 provisional ones.
8 Using only their eyes, experienced observers can just detect a variability in Saturn's brightness, depending upon the degree to which the rings are contributing.

11 Ice Giants: Uranus and Neptune

In 1738, Herschel was born into a world with six planets, but he would change all that. His 'new' planet is visible just barely to the naked eye, but easily dismissed as a dim star. Although an amateur, Herschel was armed with the most powerful telescope in 1781. He used it differently than professional astronomers used instruments: They typically pointed their 'scopes at specific, known objects. Heschel 'swept' the skies, methodically recording everything of interest he saw as the Celestial Sphere appeared to roll above him. He was rewarded when he spotted Uranus as a disk floating in the field of his eyepiece on March 13. In addition to appearing as a round shape rather than an unresolved point of light, calculations indicated that its orbit was nearly circular (unlike that of a comet, for instance).

Herschel suggested that his discovery be called *Georgium Sidus* , which is Latin for George's Star. Fellow German George III was King of England at the time.[1] However, that proposal was unpopular outside of England. Johann Bode ⟨1747–1826⟩, German astronomer and editor of a professional journal, proposed the name Uranus.[2] In Greek mythology, Uranus personifies the sky. He is also the father of Cronus, who is the father of Zeus. The Romans associated Zeus with Jupiter. This name eventually received international acceptance. In 1850, the British *Nautical Almanac* finally began using Uranus as well.

Because Uranus is theoretically just visible to the naked eye, at least 20 observations were recorded before Hershel's without the astronomer realizing the significance of what he was seeing. As more astronomers reported observing Uranus, they could refine their calculations for its position. However, discrepancies between the observations and best predictions appeared that could not be explained by interactions with any of the other six planets. Another unknown world was making its presence felt gravitationally, distorting the otherwise slightly elliptical orbit of Uranus.

Using Newton's laws of motion and gravity, Le Verrier computed the position of this new body mathematically. On 23 September 1846, German astronomer Johann Galle ⟨1812–1910⟩ detected a previously unknown 'star' within 1 degree of the location predicted by Le Verrier. Further observations confirmed this theoretical triumph. Not only could Newton's laws explain planetary motion, but they could be used to predict the presence of an unknown planet. Of course, some good luck was involved, as well; Galileo observed Neptune earlier, but moved on thinking it just another star.

A British astronomer, John Adams ⟨1819–1892⟩, had made a similar prediction earlier but with less success in initiating a search for the planet. Promoting his work, the English proposed naming the new planet, 'Oceanus.'[3] The French counter-proposed 'Neptune,' which had been previously suggested for Uranus. Ultimately, Neptune was accepted. Neptune is the Roman god of the seas.

The Ice Giants, Uranus and Neptune, are among the least explored worlds in our Solar System, and yet they represent a planetary class that may be commonplace in the Galaxy. The properties of their interiors are poorly known but may hold clues to the formation of our Planetary System. Their atmospheres represent two extremes: Uranus's is forced by unusual solar heating due to its axial tilt, and Neptune's climate is driven by a powerful source of internal energy (Figure 11.1).

Like the Gas Giants, the atmospheres of the Ice Giants are primarily hydrogen and helium. At 19 au from the Sun, sunlight takes 2 hours, 40 minutes to reach Uranus, where it is so cold that ammonia has dropped out of its upper atmosphere leaving hydrogen sulfide and methane. On Jupiter and Saturn, ammonia produces the highest distinguishable white clouds. On Uranus, methane scatters sunlight producing a distinct blue color. On the so-much-warmer Earth, we think of methane as liquified natural gas!

FIGURE 11.1   Uranus and Neptune imaged by Voyager 2.

Unlike Jupiter and Saturn, Uranus's equatorial winds move in retrograde. The high winds result in no bright zones, nor dark belts. Most of the time it looks like one big, featureless blue robin's egg. Some water clouds do appear—just barely, but only at the time of Uranus's equinoxes.

Appropriate for a sea god, Neptune has a similar methane-blue color. Additional, unknown blue compounds also may be present in Neptune's atmosphere. At 30 au from the Sun, sunlight takes 4 hours, 12 minutes to reach Neptune. The location of this world puts it closer than any other planet to the Kuiper Belt. ☞ Chapter 14 ☞ Neptune takes 165 years to complete a revolution about the Sun.

As indicated by its blue color, the composition of Uranus includes 2% methane, but it is otherwise similar to the other Jovian Planets with 83% molecular hydrogen and 15% helium. It, too, has layers of gaseous and liquid hydrogen. Uranus is not massive enough, though, to produce the gravity necessary for squeezing together a metallic hydrogen layer. Instead, planetary scientists think that Uranus's core is surrounded in onion-skin fashion by layers of different kinds of ice. Hot ice. Such a thing is possible under extreme pressures. Therefore, specialists distinguish Uranus and Neptune from Jupiter and Saturn by giving them their chill-inducing name.

Uranus may have a high-pressure carbon shell thicker than Jupiter's or Saturn's. Imagine a world-sized layer of diamond with new diamonds precipitating out of the sky! What would De Beers think?

With almost the same size and mass as Uranus, Neptune has the same compositional characteristics as the other Jovian Planets. An imaginary cross section of Neptune should look like a cross section of Uranus.

Before Voyager 2 arrived at Uranus in 1986, astronomers did not even know the planet's rotation period. Some 'high tech' image processing revealed marginal differences among the *in-situ* images of clouds. Still, the pictures were enough to finally establish a planetary rotation period

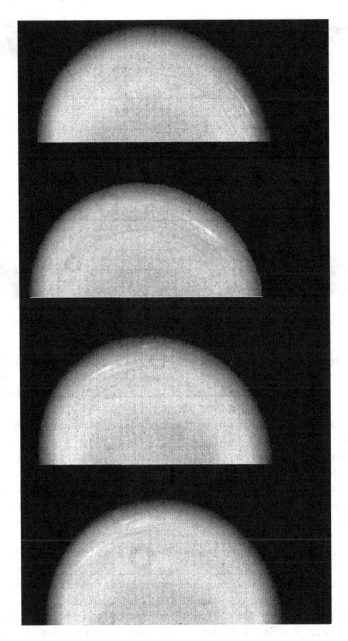

FIGURE 11.2 Voyager 2 time series of Uranus images (with contrast highly enhanced) shows the circling of clouds and allows timing the rotation period for the planet.

of 17 hours 14 minutes. Uranus rotates differentially, but the polar regions move at faster speeds circling the poles in 14 hours, unlike the other Jovian Planets (Figure 11.2).

That same level of processing revealed some color differentiation in Uranus, too: High in the atmosphere, at the poles, there is an accumulation of what looks like organic compounds. In other words, Uranus has a smog problem (Figure 11.3).

After their experience with Uranus, Voyager 2 scientists had nothing to expect in the way of atmospheric features from the further Neptune. Such weather features require an energy source. On the Earth, it is sunlight, but the Sun normally is too weak at jovian-planet distances to have much effect.

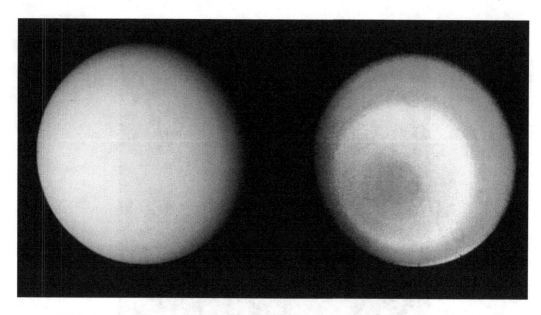

FIGURE 11.3   Voyager 2 images showing Uranus without (left) and with (right) color highly enhanced.

In the case of bulky Jupiter and Saturn, internally stored thermal energy from their formation is thought to be the culprit. But smaller Uranus and Neptune should have little remaining of this primal heat source. (Remember the small-mouthed, insulated drink bottle.) What stored energy that may have existed should have radiated away altogether by now.

Nature has a way of always surprising us. Despite receiving a minimum amount of sunlight, Neptune's atmosphere is surprisingly dynamic with large storm systems and high-speed winds. Straight-line neptunian jet streams can clock in at 2,100 kilometers per hour (1,300 miles per hour)—a solar-system record.

A fluid body, each jovian planet's apparent rotation period at different latitudes is the sum of its internal, bulk rotation and wind speed. Neptune's average rotation period is 16 hours, 1 minute, but at the planet's equator it is 18 hours. At its poles, the period is only 12 hours—a range that sets another record.

The first close-up pictures of Neptune from Voyager 2 in 1989 showed huge, unexpected, towers of white cloud jutting up through the visible atmosphere. This was true even though the *minimum* cloud-top temperature of Neptune is a mere –220°C (–370°F) (Figure 11.4).

One rotating storm on Neptune was dubbed the 'Great Dark Spot' in honor of its similarly appearing counterpart back on Jupiter. The Great Dark Spot was at approximately the same latitude on Neptune as the Great Red Spot is on Jupiter, and compared to the size of the planet, was the *bigger* feature. We use the past tense for the Great Dark Spot because it staged a jaw-dropping disappearance when HST imaged Neptune in 1994. Just as it takes energy to brew a storm like this, it takes energy to make it disappear, too. Since then, HST has observed other energetic cyclones on the planet come and go. The fact is that Neptune *radiates* energy like Jupiter and Saturn but not Uranus (Figure 11.5).

Jupiter and Saturn, like most planets with magnetic fields, have magnetic poles near their geographic poles through which run their spin axes. Not Uranus. Its magnetic field is tilted 59°—and offset from the middle of the planet by ⅓ of Uranus's radius. This odd magnetism results in aurorae far from the north and south poles. It also creates a corkscrew-shaped magnetosphere. Indeed, it has been suggested that Uranus's magnetic field switches on and off every uranian day.

Moreover, the magnetic field may precipitate Uranus actually *losing* its atmosphere: More than three decades after Voyager 2's historic encounter, astronomers analyzing the magnetic data

FIGURE 11.4   Voyager 2 image of storms on Neptune.

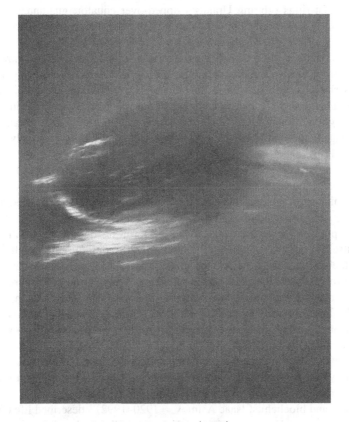

FIGURE 11.5   Voyager 2 image of the Great Dark Spot ⟨1989⟩.

discovered evidence for a giant burst of ionized gas from Uranus's atmosphere, called a plasmoid. Such events are undoubtedly most common on Uranus but are seen on other planets. Is the magnetic field the cause? Is this a case of planetary flatulence caused by indigestion from a giant collision? Maybe, but we will need to wait until a return mission to know better.

Like its fellow Ice Giant, Neptune has a significant magnetosphere that is askew, tilted from its rotation axis and offset from its center. Since neither planet is thought to contain metallic hydrogen in its interior, the source of magnetism is likely an internal shell of ionized water.

Astronomers are sometimes opportunists. In 1977, American astronomer James Elliott 〈1943–2011〉 intended to take advantage of a serendipitous occultation. An occultation occurs any time one object in the sky is covered up by another. (A total solar eclipse is an example.) In this case, it was to be the planet Uranus occulting a star. The way the point-like star disappeared behind Uranus would probe the thickness of Uranus's 'atmosphere,' as it faded behind the planet.

To make this precise measurement, Elliot would use an observatory unlike any other: a telescope mounted in the side of a jet aircraft—flying out over the ocean from where the occultation was visible and high above most of the Earth's obscuring air. The event was planned to the fraction of a second, taking into account the Earth's rotation, the plane's velocity, the position of the star, and the motion of Uranus.

But the star winked out early. And suddenly. And then again and again! Five times in all. The planetary occultation commenced on time, but once it was completed, the strange pattern of unexpected starlight disappearances recommenced. What would do that? Five new satellites lined up perfectly along the track of the occultation? Extremely unlikely.

Rings. Independent rings orbiting Uranus's tipped-over equator, guaranteed to block out the star's light—each for at least an instant—no matter what. After nearly 350 years, Saturn could no longer claim the title of *the* ringed planet.

It turns out that Uranus has two sets of rings, for a total of 13. These rings contrast with Saturn's by being very narrow and especially dark: the albedo of charcoal. The ring particles are all small—often described as 'golf-ball' size. Fine dust is 'sprinkled' among them. The rings are not quite circular, do not lie exactly in Uranus's equatorial plane, and vary in width. These irregularities are due to gravitational interactions with small nearby moons that shepherd the particles into the narrow rings in the first place. Uranus's rings are thought to be relatively young like those of Jupiter (Figure 11.6).

Neptune's rings are as meager as Uranus's. It has nine sets, the outermost of which consists of partial ring arcs concentrated into clumps, probably influenced by the gravity of small imbedded moons. The neptunian rings are unstable and slowly dissipating with an estimated lifetime of a few 100 million years. Are they resupplied? From where? (Figure 11.7)

In all the description so far, Uranus must sound like a run-of-the-mill Jovian Planet. Not so fast. It distinguishes itself in one way: its obliquity. Uranus has an obliquity of 98°. It is rolling around in the plane of the ecliptic like a barrel, the planet knocked on its side. What is more, Uranus rotates backward (clockwise), the only planet other than Venus to do so (Figure 11.8).

What screwy seasons Uranus must have! Modest obliquities produce seasons on other planets while planets with little obliquity experience minimal seasons. Every half uranian year (42 Earth years), one hemisphere of the planet is in daylight with its pole pointed more-or-less toward the Sun during the summer, the other half night, and *vice versa* . Consequently, the polar regions in summer are the hottest places on the planet. During the spring and fall, most of Uranus experiences typical day/night cycles. It is unlikely that the distant Sun would whip up even the meager cloud features seen on Uranus without its extreme obliquity and dramatic seasons.

American author and biochemist Isaac Asimov 〈1920–1992〉 described life on Uranus as: "A human being born at one of Uranus's poles would be a middle-aged man at sunset and a very old man before it was time for a second sunrise."[4] To avoid the long frigid night, a planetary migration from one hemisphere to the other would take place every 42 years. A jet setter 'on' Uranus could change her place of residence twice each year and never see the Sun go down.

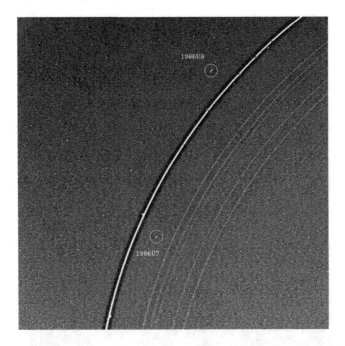

FIGURE 11.6 Uranus's rings with two accompanying shepherd satellites. ☞ Chapter 12 ☞.

FIGURE 11.7 Neptune ring arc.

With an 84-year revolution period, the seasons seem long and bizarre, that is until we look at the *minimum* cloud-top temperature of Uranus: –220°C (–370°F). Its atmosphere is the coldest of any planet in the Solar System even though it is not the farthest from the Sun. Nobody knows why. It is incredibly cold everywhere in Uranus's atmosphere, no matter *what* the season.

With an obliquity of 28° and a 'normal' rotational direction, neither the seasons nor the days on Neptune stand out. Nevertheless, Triton[5] ☞ Chapter 12 ☞, Neptune's only large satellite, provides some local color as it revolves backward around the planet. Once upon a time, this typically massive Jovian Planet may have encountered a pair of dwarf planets, ejecting one and capturing the other as a moon (Figure 11.9).

FIGURE 11.8 Diagram showing the obliquities of the planets.

So, at the edge of the Planetary System, we are left with a mystery. Uranus and Neptune should be the true 'twin planets.' Uranus is slightly larger, Neptune is slightly more massive, but their bulk properties are basically the same. They likely were formed in the same place within the Solar System at about the same time. Why then is Uranus so quiescent and Neptune so active? Why is it that Uranus, not Neptune, is the coldest member of the System?

FIGURE 11.9   Voyager 2 looking back: a crescent Neptune and Triton.

We have dealt mainly with long-term processes at work on the planets. Remember that trau-matic, extremely short-term events have a role to play, too. Foremost among these is impacts.

Maybe Uranus once had a 'normal' obliquity, like Neptune has, but was struck by a large object that hit obliquely, tipping Uranus's spin axis but barely affecting its deep interior where its magnetic field is generated. This may explain the large difference in location between Uranus's magnetic and rotational poles. More importantly, it could have released a large fraction of Uranus's small primordial heat budget all at once, resulting in the cold planet we see.

Maybe Neptune had settled into a quiet, differentiated existence, when an impactor ploughed directly into it and stirred the planet up again. Still today, it seems less centrally condensed than Uranus, and its core temperature is modest as Jovian Planets go: 4,700°C (8,500°F). The atmo-spheric activity we witness now just might be Neptune re-sorting itself into layers. Maybe.

The impact of an icy body with Uranus would answer another question: Why are Uranus's satellites all orbiting in the plane of its equator? ☞ Chapter 12 ☞ Like planets, moons most often are found closer to the plane of the ecliptic; although, in the case of low-obliquity planets, the two orientations amount to almost the same thing. Did Uranus's plentiful acolytes form all at once out of the steam produced by the impact's fireball?

Other unique planetary features also have been blamed on impacts. Why is Mercury's crust so thin compared to its core? An impact 'shelled' off its once thicker crust. Venus turns slowly and backward? An impact gave it back spin! However, just how many of these coincidental 'one off' events are we willing to accept? The planets are silent on these matters.

The astronomical community has prioritized a return mission to Uranus and Neptune in order to resolve many of these outstanding questions about the Ice Giants.

NOTES

1 George III awarded Herschel a stipend that allowed him to become a professional astronomer.
2 In pronouncing the name of this planet, put the emphasis on the first syllable.
3 In Greek mythology, the personification of the mythical river that surrounds the Earth.
4 Isaac Asimov, *The Relativity of Wrong* ⟨1988⟩.
5 Half-man, half-fish son of the Greek sea god, Poseidon.

12 Satellites of Ice and Fire

Besides the planets, dwarf planets, asteroids, and comets that all orbit the Sun, many other bodies inhabit the Solar System, namely moons, which are also known as natural satellites. Some are planet size, but they are not called planets only because they happen to orbit a planet, dwarf planet, or asteroid. If the planets are the children of the Sun, then the moons are the Sun's grandchildren. These are worthy of being called worlds in their own right.

Astronomers think that some moons formed in a disk of gas and dust encircling their host body, like the proto-solar nebula only scaled smaller. Evidence also points to other moons being captured or formed by collisions with their host (e.g., our Moon).

The eight planets host at least 219 moons, and more are found every few years. Of these, the Giant Planets hog the most moons, with Saturn and Jupiter leading the count with more than 100 between them. In contrast, the four terrestrial planets have a total of three moons. In addition, even dwarf planets and asteroids can have moons (e.g., Pluto has five). So far, no moons have been found orbiting comets (Figures 12.1 and 12.2).

Moons come in many sizes and shapes with a variety of characteristics. A few of the largest ones have atmospheres, but most do not. They are generally solid from the surface to the core, but a few are thought to have oceans hidden below their icy surfaces. The largest ones display geologically active surfaces, but most surface features have remained unchanged for billions of years. Nineteen are round due to their large size and hence strong self-gravity. The others are shaped asymmetrically. A few have intrinsic magnetic fields. Two are larger than Mercury, four are larger than our Moon, and seven are larger than Pluto. Many would certainly qualify to be planets or dwarf planets if they orbited the Sun independently (Figure 12.3).

Moons are classified according to their orbits: regular or irregular. Regular moons orbit in the direction of their planet's rotation (prograde orbits) and are near the plane of their host's equator. These moons are more likely to have formed in place around their host. Irregular moons orbit in either the direction of their planet's rotation or in the opposite direction (retrograde rotation) and often are highly inclined to their host's equator. These moons are probably asteroids or comet nuclei that have been captured from surrounding space. Most irregular moons are small, less than 10 kilometers (6 miles) in size. At about 100 kilometers (60 miles), Phoebe at Saturn holds the record as the largest known irregular moon (Figure 12.4).

Moons are usually named after mythological characters of one culture or another. An exception is Uranus. Anyone familiar with William Shakespeare's plays or Alexander Pope's poems will quickly recognize the names of several characters (e.g., Oberon, Ophelia, Titania, Puck, Belinda, Umbriel, *etc.*). When first discovered, the IAU gives an object a provisional designation that indicates the order and year of discovery. For example, S/2004 N1 was the first neptunian satellite discovered in 2004. When additional observations confirm the discovery, the IAU grants an official name. Thus, S/2004 N1 became Hippocamp[1] in 2019 (Figure 12.5).

Which was the first to be known? Our Moon, of course! Next came the discovery of four bodies orbiting Jupiter made by Galileo in 1610: Io, Europa, Ganymede, and Callisto. Clearly, these Galilean satellites did not orbit the Earth, challenging the dominant Earth-centered philosophy. This important observation revolutionized our understanding of the Universe from one where everything orbited the Earth to one centered on the Sun, as proposed earlier by Copernicus. Over the next centuries, astronomers found only a few additional moons. The detections surged with the advent of the space age and very large telescopes in the later part of the 20th century (Figure 12.6).

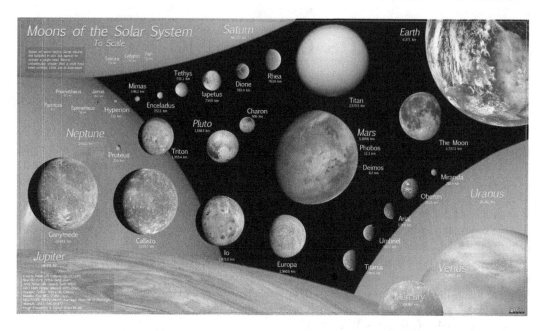

FIGURE 12.1 Selected satellites.

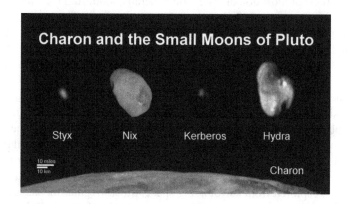

FIGURE 12.2 Montage of Pluto's five moons as imaged by New Horizons.

The Galileans hold another special place in the history of astronomy, because they were used to determine the finite speed of light. Starting in 1676, Danish astronomer Ole Rømer ⟨1644–1712⟩ noticed them apparently becoming occulted by Jupiter, or reappearing from behind Jupiter, at times other than their Newtonian orbits would predict. The problem was not with Newton; the satellites were on time. It was due to the fact that, as the distance from the Earth to Jupiter and its moons varied, it took more or less time for the light signaling these events to reach us. Using Rømer's timings, Huygens found a value of 210,000 kilometers per second (130,000 miles per second), about 30% lower than the correct value.

We will start our tour of the satellites of the outer Planetary System with the four Galilean moons, orbiting from 1.1 million kilometers (660,000 miles) to 420,000 kilometers (260,000 miles) from Jupiter. They are all in regular, prograde orbits and rotate synchronously with Jupiter, always keeping the same hemisphere facing the planet like our Moon; they probably formed along with their planet. Their sizes range from Ganymede, which is larger than Mercury, to Europa, which is slightly smaller than our Moon. These moons are not small worlds but only appear that way when compared in

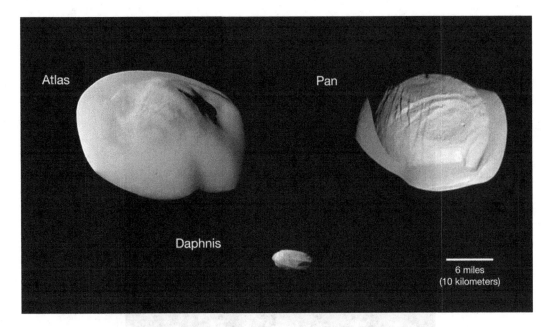

FIGURE 12.3 Cassini orbiter images of small satellites. One looks like Saturn with its rings!

FIGURE 12.4 Orbits of Saturn's irregular satellites.

diameter to the behemoth planet they orbit. If they were not lost in the glare of nearby Jupiter, you could see the Galilean satellites with your naked eye (Figure 12.7).

We visit the Galilean satellites individually in order of decreasing distance from Jupiter. They are named, by the way, after lovers of the god Jupiter in Roman mythology.

The first is Callisto, the second largest, with a density of 1.8 g/cm^3. This density is representative of most similar outer-solar-system satellites; it is between the 4 g/cm^3 for the terrestrial planets and the 1 g/cm^3 or so for the Jovian Planets. This tips us off that there is something fundamentally different here. These satellites are neither terrestrial nor jovian. What gives us a

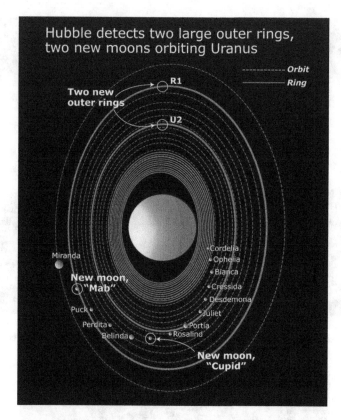

FIGURE 12.5 Satellites of Uranus, including two new ones discovered using HST.

FIGURE 12.6 Galilean satellites may be seen through even an amateur telescope.

density of about 2 g/cm³? A mix of ice and rock does so nicely. Callisto is thought to be differentiated just this way: about 50% water ice surrounding a rocky core. Hence their classification as Icy Satellites, a third kind of solar-system body. But dirty ice—nothing you would want to find floating in your soft-drink glass! (Figure 12.8)

Ice is usually thought of as bright, but the surface of Callisto is not. One reason is dirty ice and another is the multitude of crater-forming meteoroids that have struck the satellite. Its surface is

FIGURE 12.7   Galilean satellites to scale.

FIGURE 12.8   Galileo close-up of Callisto.

covered by fine, dark material. Callisto's surface is old enough to be completely saturated with craters; it is perhaps billions of years old, one of the oldest, most heavily cratered in the Planetary System. This prevents us from trying to date the surface by counting the number of craters with which it is marked. If you want to add a crater to Callisto, you have to remove one to make room!

Craters formed in viscous ice, like that on Callisto, look a bit different from those on terrestrial surfaces. The ice at the bottom of the crater eventually relaxes, leaving a dome-like structure surrounded by the crater rim. Truly big craters are hard to find on icy surfaces; the higher the structures they create (e.g., mountains), the quicker these formations will flow to surface level. Really old craters have retained no relief at all and are simply circular markings on the ground—like a palimpsest.

Callisto has eight sets of craters arranged in a straight line known as a catena,[2] or more commonly, a crater chain; they are named from Norse mythology. Astronomers think catenae formed when an approaching impactor, likely a comet, fragmented into a string of smaller bodies due to tidal disruption. ☛ Chapter 13 ☛ Gipul³ Catena is its longest at about 640 kilometers (400 miles) in length, with individual craters up to 40 kilometers (25 miles) across.

Although Callisto boasts the most catenae in the Solar System, the Moon, Mars, Ceres, and other satellites of the Jovian Planets also are marred by catenae. In addition, secondary impacts from crater ejecta, tectonic activity, and volcanic activity can also produce catenae.

Callisto has a magnetic field, most likely generated by electrical currents in a deep subsurface ocean! Electrically conducting, briny water responds to Jupiter's powerful field as it sweeps by, generating magnetism that varies with time. The hidden sea is deep enough to not affect Callisto's ancient cratered surface.

Callisto is a dead world, geologically inactive. Maybe it never had such activity. It just progresses around Jupiter, taking it on the chin, battered by meteoroids.

Ganymede orbits Jupiter in resonance with Europa and Io. It is the largest satellite in the Solar System and, coincidentally, has a male name. Ganymede is only a little denser than Callisto (a bit more dirt, a bit less ice). This giant satellite is cratered like Callisto, but not uniformly. There are two distinct types of terrain on Ganymede: One-third of its surface is covered with dark polygons having more-or-less straight edges; in-between, there are broad, lighter stripes. We know Ganymede's surface is ice; the lighter regions must be cleaner ice. Because these regions have accumulated fewer craters than the polygons, we reason that they are newer. There are also regular grooves in this terrain, parallel to the polygon edges.

What is going on here? The grooves are thought to be nothing less than stretch marks on Ganymede. At one time, the surface of the satellite may have expanded upon freezing, just as an ice cube overfills its tray when water becomes frozen. The polygons are Ganymede's old crust pulled apart. The lighter material welled up later to fill the gaps. Ganymede must once have experienced geological activity. It simply does not appear to be doing so any more (Figure 12.9).

Ganymede has an intrinsic magnetic field, making it the only satellite that currently generates its own magnetism. This is an interesting case of a magnetic bubble nestled within Jupiter's much larger magnetosphere. Ganymede is a fully differentiated body with an iron-rich, liquid core and a buried sea that may contain more water than all of the Earth's oceans combined. Like that of the Earth, the magnetic field likely is caused by electrical currents that swirl in its conductive core.

The next Galilean satellite down is Europa. It is not as big as Ganymede, but it is denser. It has a high albedo.

Having no mountains or valleys and few impact craters to mar its surface, Europa looks like a glass paperweight. Indeed, it is the smoothest known object in the Solar System. The only other visible feature on Europa is a pattern of crisscrossed cracks. Astronomers are looking at bright, clean, new (salty!) ice.

As the satellite differentiated, water came to the top. Lots of it. A comparatively warm aqueous layer formed all over Europa, with a frozen crust over it. Europa may be the only world in the Solar System with a global ocean—besides the Earth.

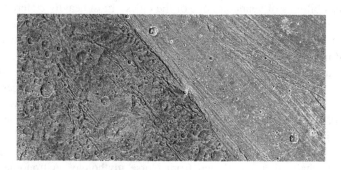

FIGURE 12.9   Galileo close-up of Ganymede.

Jupiter-Europa tides cause cracks in the exterior ice. Water released to the surface has smoothed over old craters. Since new craters have not accumulated to roughen Europa up, its geologic activity must have ended only recently or may still continue.

What is the depth of the top ice? It may not be thick. The cracks border larger blocks of ice that appear to float on Europa's surface. Close-up pictures of the satellite resemble those taken of the Earth's Arctic in winter, when thin ice overlays deep ocean (Figure 12.10).

Scientists speculate that Europa's ocean could be a suitable environment for the development of life, having the necessary ingredients: water, likely organic material, and an energy source provided by tidal friction discussed below. If we were to send a submarine to Europa, break through the ice, and then descend, what would this aqua-craft see? Recall that astronomers think other outer-solar-system satellites, such as Callisto and Ganymede, have some briny subsurface water, too.

Like Callisto, Europa has a magnetic field. In the case of Europa, it is induced by Jupiter's powerful field sweeping by, which apparently generates electrical currents in a subsurface ocean.

The innermost Galilean satellite is Io (sometimes pronounced 'E-oh,' definitely not 'ten'). Io is close enough to giant Jupiter that a vast current of electricity flows between it and the poles of Jupiter. The enormous power generated produces aurorae on both worlds.

Io is a little bigger than our Moon and is the densest of the four Galilean satellites. The first pictures of Io from the Voyagers arrived back at the Earth revealed it to be without-a-doubt geologically active; it is in fact the most active in the Solar System.

Around the edge of the satellite's limb, Voyager captured active volcanoes, throwing plumes of gas as much as 500 kilometers (310 miles) against the low gravity of Io, which then gracefully

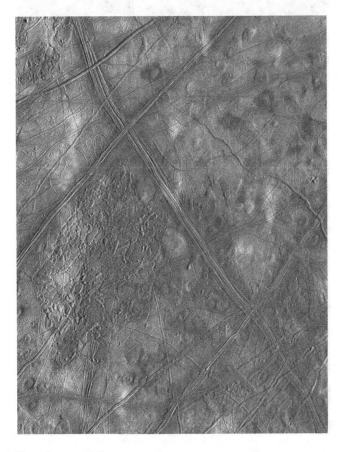

FIGURE 12.10   Galileo close-up of Europa.

descend umbrella-like and freeze onto Io's surface as a white frost. Lava flows create plains on its surface among more than a hundred volcanic mountain ranges, some taller than Mount Everest. This volcanism recycles interior material to the surface and effectively turns Io inside-out. Io is the only world, besides the Earth, to have volcanic eruptions taking place today, excluding cryo-volcanism discussed below. In fact, the word 'eruption' may be misleading since many of Io's volcanoes have been constantly spewing material since they were discovered (Figure 12.11).

We see a surface of Io spotted with black, yellow, red, brown, white, and even blue. These are the colors of the element sulfur exposed to different temperatures. Sulfur and sulfur-rich organic compounds escape in plumes from the volcanoes and cover the surface. Volcanoes are filled with melted silicate rocks that are hotter than any planetary surface, even Venus. Picture Io as a rocky core surrounded by a thin layer of sulfur. The impact crater count on Io is nil. Craters are covered over as soon as they are created. Mapping Io is impossible because its face is forever changing. In truth, calling Io an Icy Satellite at all is inappropriate. Its principal, observable ingredients are rock and sulfur. It is a moon of fire!

As Jupiter's magnetic field sweeps past Io, it picks up thousands of kilograms of sulfur and oxygen ions, directing them into a doughnut-shaped ring known as a plasma torus, centered on Io's orbit. Indeed, the key to most of what astronomers see going on with Io is sulfur. Io even leaks sulfur: A small satellite (captured asteroid?) named Amalthea orbits Jupiter close to Io. It has an unusual orange tinge. It is being dumped upon by Io!

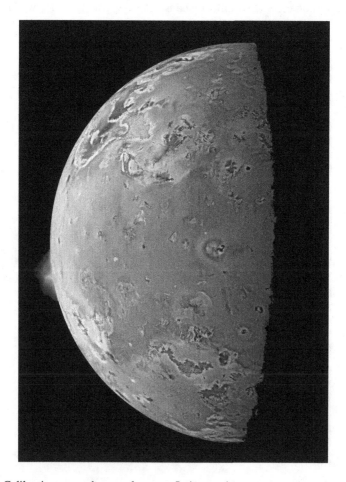

FIGURE 12.11   Galileo image catches a volcano on Io in eruption.

Where does all the heat energy come from to drive Io's geologic activity? Given its size and mass, Io could not have held onto a primal heat supply. It is certainly too far from the Sun to collect meaningful amounts of solar energy. The answer seems to be that Io is an energy thief, and it preys upon Jupiter.

In addition to being so close to Jupiter, Io also has an eccentric orbit such that the gravitational pull of Jupiter, along with that of its neighboring satellites, is always changing. These forces squeeze and stretch Io, creating and dissipating a 100-meter (330-feet) tidal bulge in its solid body. In the process, internal friction inside Io—the rubbing of rock against itself—heats up the satellite, making it molten inside and producing volcanoes. (Try rubbing your hands together: They get warm, do they not?) It is almost strange that we end with a satellite with such a short name being so complex.

Let us now take a quick look at the satellites of Saturn. In doing so we skip over the many captured asteroids and other small satellites that orbit all the Jovian Planets.

Starting Saturn's Icy Satellites with the outermost and moving inward, we first come to Iapetus. Iapetus appears to be a geologically dead world. A dark feature covering nearly one hemisphere of Iapetus appears to be some sort of organic goo. A mountain ridge surrounds Iapetus's equator. Some suspect it represents a ring that once encircled Iapetus but fell to the surface below (Figure 12.12).

Titan! We must spend a little more time with Titan because it is so unusual. It is the only solar-system satellite with an appreciable atmosphere. Its atmosphere consists of nitrogen and is at least 50% denser than the Earth's. A layer of chemical 'smog' nearly enshrouds the atmosphere, rising and descending with the seasons. These conditions blocked our view of the ground until the Cassini spacecraft released a probe, Huygens, that parachuted through Titan's atmosphere.

Huygens revealed a fascinating and uniquely Earth-like topography: choppy seas, an island, lakes, dry lake beds, branching rivers, canyons, sink holes, and dunes. However, water would be frozen and stronger than concrete on Titan's surface. Planetary scientists are not seeing a water cycle here, but rather a multistate cycle of organic compounds such as methane and ethane. There are hints of this material being stirred up into seasonal, tropospheric storms. Beneath these surface features, an underground ocean of water still may lie.

With all this methane, the greenhouse effect warms Titan above otherwise cruel saturnian temperatures. Make no mistake; it is still cold on Titan. Still, would it not be ironic if Titan were to be a more hospitable place for human exploration than the nearer, Sun-scorched crust of Mercury; oppressive surface of Venus; or even vacuum-exposed façade of our own Moon (Figure 12.13).

FIGURE 12.12 Cassini image of Iapetus.

The rest of Saturn's larger satellites are typical icy worlds. Rhea, the second largest saturnian satellite, has a conspicuously low density, suggesting it is mostly ice. An extremely thin atmosphere of oxygen encases it. In addition, it is the only satellite to have a ring.

Rhea's two hemispheres are markedly different. One is heavily cratered, the other not so. Many outer-solar-system satellites behave like our Moon: Their rotation and revolution periods are the same. However, having a nearside and a farside also means that a satellite must have a leading hemisphere and a trailing hemisphere. Rhea's leading hemisphere is scarred by the most craters as it protects the trailing hemisphere like a shield.

Dione is the next Icy Satellite to which we come. It, too, has a synchronous rotation period. This far from the Sun, an Icy Satellite is not just water ice. Other substances freeze at saturnian temperatures such as carbon dioxide, ammonia, and other volatiles (Figure 12.14).

Tethys is the least dense saturnian satellite. It is dominated by a huge crater with a central peak, Odysseus, which is two-fifths the diameter of the satellite itself! A gigantic crack radiates from this crater, presumably caused by the same impact. We are reminded that ice is more brittle than rock (Figure 12.15).

Once we get to Enceladus, we begin to see signs of significant geologic activity. This world appears to be spurting warm water and organic molecules out of great cracks in its surface, particularly at the south pole. Much of this matter snows back down. Indeed, Enceladus has one of the highest albedos in the Solar System. In addition, the rings of Saturn are brighter near this moon, as the satellite graciously provides a fresh ice coating to the ring particles. This material is also the source of Saturn's outer E-ring (Figure 12.16).

Enceladus is the smallest geologically active satellite of which astronomers know. Along with salt water, its fountains spew materials we associate with the bottom of a global ocean. Are there hydrothermal vents present here, the plumbing of which reaches far into deeper and warmer Enceladus? If so, the heating mechanism would be similar to that driving activity on Io. On the Earth, geothermal vents have developed entire deep-sea ecosystems. In fact, if we are ever to find life in the Solar System some place beyond the Earth, Enceladus is a prime candidate.

FIGURE 12.13 Huygens view of Titan's surface.

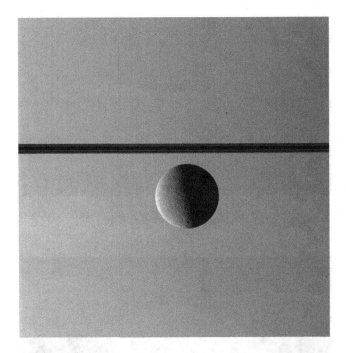

FIGURE 12.14   Cassini image of Dione in front of Saturn. The rings appear nearby.

FIGURE 12.15   Cassini image of Tethys.

Tiny Mimas has a crater, Herschel, in its side, 1/3 the diameter of the satellite itself. This gives Mimas the nickname the 'death star' among *Star Wars* fans. If what hit Mimas were any bigger, there might not be a Mimas at all today (Figure 12.17).

Mimas's rotation period and revolution period around Saturn are more-or-less the same (not surprising), but they are not exactly the same (surprising). This may be the result of a non-spherical

FIGURE 12.16 Cassini image of Enceladus showing 'geysers.'

FIGURE 12.17 Cassini image of Mimas.

distribution of mass within Mimas. Mimas might feature an American football-shaped core. Should Mimas itself then not have this shape? Maybe the core and mantle/surface are physically decoupled from one another by a global ocean.

Among Uranus's 27 known satellites, the regular ones may have formed from the same giant impact presumed to be responsible for Uranus's unique axial tilt. Cupid is the smallest at only 18 kilometers (11 miles) across. The major ones, Oberon, Titania, Umbriel, and Ariel, are classic Icy Satellites, some of which show signs of possible geologic activity.

Miranda would be a typical Icy Satellite, except it does not have a totally spherical form. It looks like a moon that was broken into fragments during an impact. Then, gravity reassembled the shards. However, not every piece of the puzzle ended up in the right place, creating a bizarre and exotic tableau! Whether this really happened is controversial. Regardless, Miranda looks like a Frankenstein satellite (Figure 12.18).

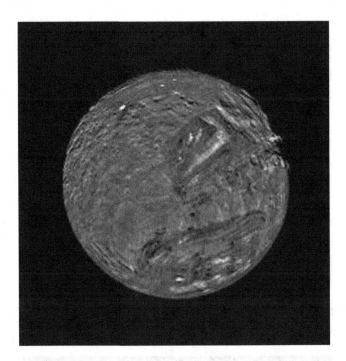

FIGURE 12.18   Voyager image of Miranda.

Pretend that you are standing on Miranda, which boasts the greatest variation in topography anywhere in the Solar System, including its highest cliff, Verona[4] Rupes. Imagine looking over the edge of this 20-kilometer tall cliff. Imagine falling off this 20-kilometer tall cliff! You would have plenty of time to assess the advisability of your action because, in Miranda's low gravity, you would take 12 minutes to fall. Then, you fire your jet pack to avoid hitting the ground at 200 kilometers per hour (120 miles per hour), stand up, brush yourself off, and, return to do it again! Miranda could be the grandest amusement park in the Solar System.

One more big Icy Satellite in the Planetary System exists, which was discovered only in 1846. An irregular satellite, Triton[5] has a highly inclined, retrograde orbit around Neptune. One of 14 known neptunian moons, Triton is the largest at 2,700 kilometers (1,700 miles) in diameter; it is only slightly smaller than Europa. Triton has a tenuous nitrogen atmosphere with trace amounts of methane and carbon dioxide.

A surprise awaits us here. For Triton has geyser-like features spread over its surface. We already have explained why geologic activity is not expected on these distant satellites, but the images do not lie. The cause of the geysers is unknown, another mystery at the edge of the Planetary System. ☞ Chapter 13 ☞

This activity falls under the general term of cryovolcanism. Cryovolcanism differs from 'regular' volcanism in that it is a much shallower phenomenon, occurs at a much lower temperature, and expels a 'cryolava' of liquid volatiles like water, ammonia, and methane. Some refer to this phenomenon as 'ice geysers' (Figure 12.19).

With more than 200 moons in the Solar System, we cannot describe all in detail. So, we end this chapter with a potpourri of others having interesting characteristics.

For instance, Epimetheus and Janus are tidally locked, sister moons of Saturn that are co-orbital (that is, they share the same orbit). Epimetheus orbits a bit faster, and, as it approaches Janus, the two switch places. These tiny dancers perform this intricate swap once every four years.

Saturn's Hyperion has a spongy appearance and chaotic rotation that stands out among an already strange collection of satellites. It also carries an electrical charge that shocked the Cassini orbiter as it sped by (Figure 12.20).

FIGURE 12.19 Triton's appearance has been likened to that of a cantaloupe! The dark spots indicate sites of cryovolcanism.

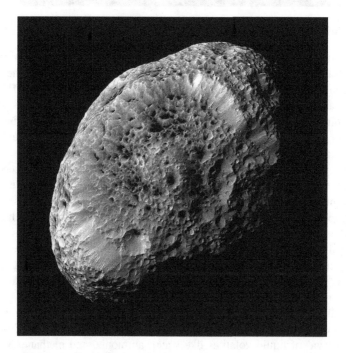

FIGURE 12.20 Cassini image of Hyperion.

Pan and Atlas are small moons found in the gaps of Saturn's rings; they are each about 30 kilometers (18 miles) and shepherd ring particles, keeping the ring gaps open. They acquire ring material that accumulates at their equators, resulting in appearances that resemble flying saucers or ravioli, your choice (Figure 12.3).

Daphnis is a tiny moon located in the gap between Saturn's A and B rings. It maintains the division and causes waves of ring particles at the gap's edge. Daphnis also accumulates ring particles around its equator, causing a narrow ridge.

Neptune's innermost moons, Naiad and Thalassa, have a strange orbital resonance that causes them to do-si-do. Naiad appears to 'pirouette' around Thalassa's orbit every 22 days in order to avoid collision.

Now we return to small bodies in the Solar System: icy interlopers coming from the farthest depths of the Solar System known as comets.

NOTES

1 A mythological seahorse in Greek mythology, a symbol of Poseidon.
2 Latin for 'chain,' plural catenae.
3 Gipul was one of the rivers which ran from Asgard to Midgard.
4 Italian city where Romeo and Juliet lived.
5 Not to be confused with Titan, back at Saturn; Saturn is a Titan while Triton is a son of Neptune.

13 Small Bodies: Comets

Comets are strange beasts. They have frightened and fascinated people since earliest historical records, dating back to the 2nd millennium BCE. Comets appear unannounced in the night sky, contrary to the predictable motions of other celestial bodies (e.g., Sun, Moon, and planets). Most move along paths different than any planet. They do not *look* like anything else, with their fuzzy heads and long tails. In fact, *comet* derives from a Greek word meaning 'hairy' star, due to the resemblance of the tail to streaming hair. Occasionally a Great Comet cometh, being seen in broad daylight and rivaling the brightness of the Moon! Then, after weeks or months, they disappear from our skies (Figure 13.1).

These mysteries and the unpredictability of comets caused them to be considered omens of terrible goings on in the world. Their sword-like appearance reinforced this view. Without scientific thought, ancient cultures often turned to myth and legend to explain them. Astrologers, soothsayers, and charlatans used comets to ply their craft. Sadly, as there is something amiss in the world much of the time, these harbingers often appeared to be right (Figure 13.2).

During the Renaissance, some coins and medals displayed comets. These items were used as amulets to ward off the bad luck from the comet. Even as late as the 18th and 19th centuries, people believed that comets were a sign that the world was coming to an end. And in 1997, the Heaven's Gate cult in San Diego, California, committed mass suicide as Comet Hale-Bopp approached, believing that the comet was harboring a spaceship that would take them to the afterlife. As we shall see, the modern view of comets stands in contrast to these ancient beliefs and superstitions.

The Renaissance heralded the beginning of scientific investigations into comets. Aristotle argued that comets are phenomena entirely within the Earth's atmosphere. Meteors were definitely seen to be within the atmosphere, strengthening their connection to comets at the time. Brahe established their celestial nature. He did this by comparing the simultaneous positions of a comet relative to the background stars from several locations on the Earth. A nearby object will appear to shift its position when viewed from these different perspectives, called parallax. Brahe reasoned that comets must be farther from the Earth than the Moon.

Newton suggested that comets follow all the rules that planets do. If it was true, it would take away some of the mystique about comets. In 1696, Halley[1] noticed, in historical records, reports of bright comets in the years 1531, 1607, and 1682. That was once every 76 years. Maybe it was the *same* comet, orbiting the Sun, and passing near the Earth where we could see it, in that interval of time. Using Newton's new theory of orbital motion, Halley stuck his neck out and predicted the return of the comet in the year 1759. Halley was not around anymore to see it, but the comet did return faithfully that year and has done so about every 76 years since, most recently in 1986[2] (Figure 13.3).

Now, we have just explained a popular misconception about Halley's Comet. Halley did not discover 'his' comet. Searches of ancient records reveal sightings as early as 240 BCE by Chinese astronomers. However, until Halley, everybody thought it was a different comet each apparition. Halley did something more important than discover a single comet. He demonstrated that comets are part of the Solar System, orbiting the Sun much like planets (Figure 13.4).

There is reason for confusing Edmund Halley as the discoverer of his eponymous comet. Ever since, when a new comet becomes visible in our sky, it is named after its discoverers. German-English astronomer Caroline Herschel ⟨1750–1848⟩ was one such comet hunter, having discovered several comets. Another prolific comet discoverer is American astronomer Carolyn Shoemaker, with 32 comets and over 800 asteroids to her credit. Today, many amateur astronomers scan the night sky

FIGURE 13.1   Comet West (1976).

every evening with their telescopes for a glimpse of a new, fuzzy patch, a newly recognized comet, and their slice of immortality.

Predicting when the next bright comet will appear in our sky is impossible. In 1973, astronomers were excited about the discovery of a new comet. Their estimates showed that this comet should become quite bright in our sky. Books were written, and television programs were broadcast preparing us for the spectacle of the comet. Alas, Comet Kohoutek barely could be seen with the naked eye. People felt as if they had been fooled, and so, when a truly great comet appeared overhead in 1976, many people ignored it. They therefore missed Comet West, one of the most beautiful comets to this day. Even now, astronomers do not understand all the factors that make a comet bright and showy.

While the public was disappointed, Kohoutek rewarded the scientific community with many new results. These included the first observations of a comet aboard a manned spacecraft, the orbiting Skylab station, and the discovery of hydrogen cyanide, solving a 150-year old riddle of the origin of cyanide in comets.

After these comets of the 1970s, there was a long wait. The next naked-eye comet show did not appear until the 1990s. And there were two of them! Comet Hyakutake was a small comet, but it came close enough to the Earth that it spread its tail half-way across our sky. Comet Hale-Bopp was discovered while still far away; therefore, it was a large comet and one that did not disappoint. It has the record for being the longest-observed comet without optical aid, at 1½ years (Figure 13.5).

Why do we study comets or, rather, comet nuclei? 'Comet' refers to the apparition as it appears in the sky, complete with its head and tails, as opposed to 'comet nucleus,' which is a small icy planetesimal and the source of all comet activity. Either way, these small bodies are left-over remnants from the formation of the Solar System.

FIGURE 13.2 Bayeux Tapestry. King Harold gets the bad news about the Norman invasion. Comet Halley hovers overhead.

Having formed far from the Sun and been hibernating in cold storage ever since, comet nuclei preserve important information. In particular, they can reveal physical conditions and chemical composition of the Solar System at the time and place of their formation. Instead of bearing bad tidings, comet nuclei are messengers that bring essential clues about the origins of our Solar System. Studying them is like doing 'archeology' of the Solar System, digging back to its earliest days.

Astrobiologists think comet nuclei may provide insight into the origins of life on the Earth. They are composed of volatiles, mostly water ice with simple organic molecules; the same mixture is also essential to the prebiotic chemistry necessary for life. Since many comets cross the Earth's orbit, collisions are inevitable; estimates indicate such events occurred frequently during the Earth's early history. Thus, comets may have imported the ingredients necessary for the primordial soup conducive to living systems; molecular panspermia is the theory that comets and asteroids spread the building blocks of life to planets.

Once again, whatever collisions can provide, they also can take away. A cometary impact can have devastating global effects on life. Although such events may have seeded life, extinction events through geological time have been mapped to comets or asteroids crashing into the Earth. Recall that many paleontologists blame the demise of the dinosaurs on either a comet or an asteroid strike. ➤ Chapter 9 ➤

Astronomers have observed thousands of comets. Based on their orbits, comets belong to two categories: short period and long period. Short-period comets still may have revolution periods up to about 200 years; they often are referred to as periodic comets. Long-period comets, which make up about 3/4 of those known, have periods on the order of a million years. From where do comets come? This difference in periods suggests two separate reservoirs of comets: the Kuiper Belt and Oort Cloud. ☞ Chapter 14 ☞

Most short-period comets have very eccentric orbits. While the orbits of most planets are nearly circular, a cometary orbit definitely looks like an ellipse. Short-period comets usually travel as far

FIGURE 13.3 Oil painting of Edmund Halley.

from the Sun as the orbits of Jupiter and Saturn, but some venture out to Neptune. Their perihelion may be closer to the Sun than the Earth; some, known as sun-grazers, even approach the Sun closer than Mercury. Other than the eccentricity of their orbits, short-period comets generally behave like planets: Their prograde orbits lie near the plane of the ecliptic, and they shine by reflected sunlight. With its retrograde orbit, Halley's Comet is an exception.

Short-period comets can be further grouped as Encke-type comets [ETCs], Jupiter-family comets [JFCs], and Halley-type comets [HTCs] based on their orbital period. The ETCs are named after Comet Encke. With periods of a few years, ETCs are confined to the inner Solar System with orbits that do not reach as far as Jupiter. With periods generally less than 20 years, JFCs have been strongly affected by Jupiter's gravity; their orbits have been reduced to a size and inclination similar to that of Jupiter. With periods up to about 200 years, HTCs have orbits that can extend beyond Neptune's orbit and are inclined at random angles to the ecliptic. About 1,000 short-period comets are known, the largest group of which are the JFCs with almost 700 members (Figure 13.6).

Some short-period comets are classified as Centaurs,[3] another interesting group of small bodies, orbiting between Jupiter and Neptune near the ecliptic. Five hundred or so are known. They may display characteristics of both asteroids and comets (hence their name). Many show cometary activity, and some have been given comet designations (e.g., Chiron, Echeclus, and Oterma). But others do not. At their distances from the Sun, it is too cold for water sublimation to drive activity, and other volatiles, such as carbon dioxide and carbon monoxide, are difficult to detect observationally.

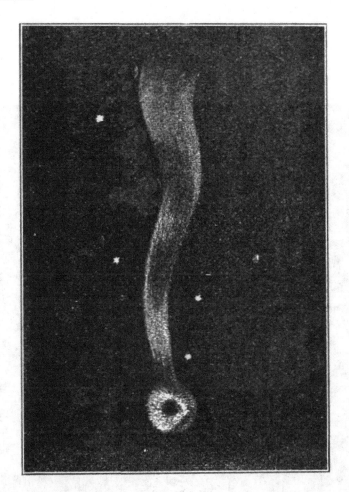

FIGURE 13.4   Comet Halley (1682).

FIGURE 13.5   Comet Hyakutake (1996; left); Comet Hale-Bopp (1997; right).

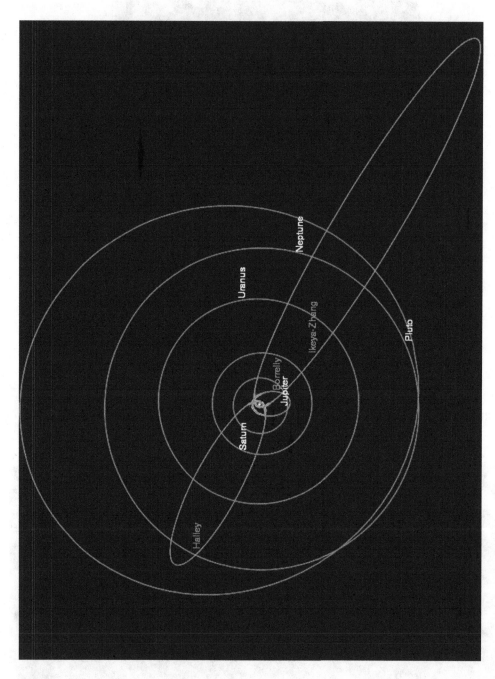

FIGURE 13.6   Orbits of three short-period comets.

The largest Centaurs are a few hundred kilometers (miles) in diameter, giving them greater gravitational powers than the typical comet. Astronomers have found two sporting rings: Chiron[4] and Chariklo![5]

These icy rogues have unstable orbits with lifetimes of a few million years. Therefore, the Centaurs may transition to the inner Planetary System in the future, possibly becoming JFCs. Astronomers predict that Comet ATLAS, a Centaur discovered in 2019, will embark on this inward journey in the next 40 to 50 years, allowing them to monitor closely its transformation to a short-period comet.

As for the origin of Centaurs, astronomers think that they came from the Kuiper Belt ☞ Chapter 14 ☞, providing a bridge to the inner Planetary System, although some Centaurs even may be interstellar visitors. So far, no spacecraft have visited any of the Centaurs, but they are certainly worthy of a future probe.

Long-period comets have *extremely* eccentric orbits. They approach the Sun as close as or closer than short-period comets, but reach aphelion far, far beyond Neptune. How far away do they get? Among those that have had their orbit calculated, some travel 1/3 of the way to the nearest star! Correspondingly, long-period comets have orbital periods measured in thousands or even millions of years. The orbital inclinations of long-period comets are random as are their orbital directions, prograde versus retrograde. Consequently, we can see a long-period comet come from any region of the sky, which is part of their unsettling 'personae.' Obviously, many more unknown examples exist because they have not gotten around to visiting the inner Planetary System since humans huddled in caves. Whereas short-period comets have encountered the Sun numerous times, some long-period comets are making their maiden voyage to the inner Planetary System (Figure 13.7).

Since comets are relatively low-mass bodies, the gravity of more-massive planets affects them strongly. Their orbital periods can vary due to encounters along their way. For example, when passing close to Jupiter, a comet can slow and arrive late to the vicinity of the Earth.

Such interactions can convert a long-period comet to a short-period one. As the gravity of the massive planet drags the comet's aphelion inward, the comet will have more frequent opportunities to encounter other planets as well. These events will eventually shorten the comet's period and reduce the inclination of its orbit into the ecliptic. Voilà! A short-period comet.

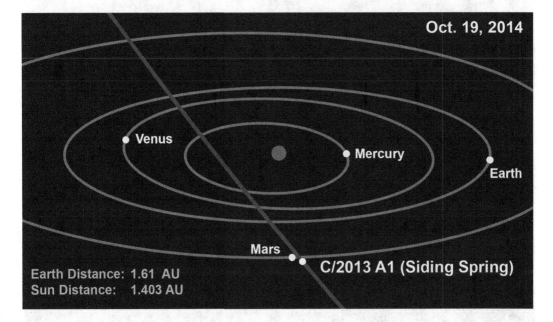

FIGURE 13.7 Orbit of long-period Comet Siding Springs (2014). Notice that its path is so eccentric, it is difficult to discern a curve.

Alternatively, such interactions can eject a comet from the Solar System altogether, sending it traveling into interstellar space. In which case, it would be known as a hyperbolic comet.

In 2017, astronomers tracked the first known interstellar object, Oumuamua,[6] arriving at the Planetary System on a path that was clearly not bound to the Sun *at all*. Then, in 2019, the Ukrainian amateur astronomer Gennadiy Borisov spotted another interstellar interloper, this time with a tail, which now bears his name. Interstellar comets appear to be as common as theoreticians have suggested all along. Through these exciting discoveries, we study material from outside our Solar System without having to travel there! Early results show that these comets are very similar to our comets, including their general chemical makeup. With the short-period and long-period comets, interstellar comets may be considered a third class (Figure 13.8).

So far, we have not written much about the comet itself—just its motion. The comet might as well have been a croquet ball. However, it is much more interesting than that.

What really happens when a comet approaches the Sun? Far from the Sun, a comet is a small body at most tens of kilometers in diameter. No wonder we cannot see this nucleus from the Earth. According to Kepler's Second Law, a comet moves at its slowest when near aphelion, so most of the period of a comet is spent beyond our view. However, when the comet comes within approximately 3 au from the Sun, solar heat begins to sublimate the ice. In the cold vacuum of space, water ice changes directly to gas when heated, a phase change called sublimation. The behavior is like that of a block of frozen carbon dioxide, or 'dry ice,' on the Earth. A coma, or bright, spherical cloud of gas and dust, forms around the comet nucleus.

Under a comet's low gravity, gas freely escapes, carrying dust with it. The coma keeps getting bigger, constantly being fed by the sublimating nucleus. The comet with coma may exceed the size of Jupiter or even the Sun, extending outward to several hundred thousands of kilometers in all directions. Near the Sun, solar ultraviolet rays excite cometary atoms and molecules so that they emit visible light, like a poster that glows under a blacklight. Scientists can identify the chemical

FIGURE 13.8   Artist's impression of Oumuamua.

composition of the coma from its signature in the fluorescence. Along with a broad envelope of dusty particles to reflect sunlight, the coma becomes visible from the Earth. The roughly spherical coma can be thought of as a fountain of gas and dust emitted by the nucleus that is eventually pushed away from the Sun by sunlight (Figure 13.9).

Molecules fragment when exposed to sunlight, so determining what chemical species are present in a comet nucleus based on its coma is complicated. What we observe in the coma is not necessarily what existed in the comet nucleus. We must infer. Laboratory experiments are essential for this detective work, such as those of African-American William Jackson, a leading astrochemist in this field. When water fragments, hydrogen is released and travels far from the coma at distances of millions of kilometers. This forms another common attribute of comets, the spherical hydrogen envelope, unseen by the naked eye (Figure 13.10).

As a comet approaches the Sun, it produces its third and most impressive feature: its tail(s). This fan-shaped appendage may grow to 1 au or more in length and become, temporarily, the largest entity in the Solar System. Traditionally, comet tails come in two styles: a gas, or ion, tail and a dust tail (Figure 13.11).

The ion tail has a bluish tint and streams straight away from the Sun. Although ancient Chinese astronomers recorded this phenomenon in 635 BCE, it was not understood until 1951. In that year, German astronomer Ludwig Biermann ⟨1907–1986⟩ suggested a stream of particles from the Sun, the solar wind, was responsible for the tail pointing anti-sunward. ☞ Chapter 15 ☞ Similarly, the flag on a ship always points away from the wind no matter which direction the ship sails.

In 1957, the Swedish physicist, Hannes Alfvén ⟨1908–1995⟩,[7] explained how ion tails form. Cometary ions present an obstacle to the solar wind. The magnetic field carried by the solar wind drapes around this obstacle, stacking up in front of and bending around the comet in the tail direction. The ions follow this draped field, forming the ion tail.

Occasionally, the ion tail separates from the comet and streams away in the anti-sunward direction. A complex interaction between the cometary ions and the constantly changing solar wind causes these disconnection events. Like a lizard, the comet grows its tail back after a short time.

FIGURE 13.9   Coma of Comet Hartley 2 imaged by the Deep Impact ('DIXI') space probe (2010).

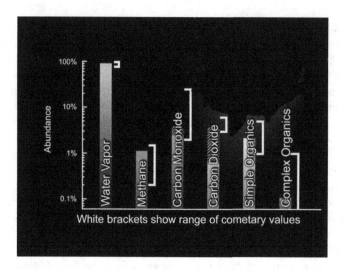

FIGURE 13.10   Gases typically found in the coma of a comet.

Due to the ion tail's interaction with the Sun's wind, comets can be used to probe the solar wind. Much has been learned about its nature using this technique, particularly at high ecliptic latitudes where spacecraft usually never venture.

The dust tail shines by reflected sunlight from each individual particle, resulting in its yellowish hue. During sublimation, dust and bits of rock are dislodged from the nucleus. The dust grains are entrained in the gas flow but leave the nucleus more sluggishly because each is much more massive than a gas molecule. The slight pressure of sunlight shapes the dust tail into a large arcing fan, in contrast to the straight ion tail. Although a dust tail can stretch across a large portion of the sky, the significance of the dust tail changes as the amount of dust varies from comet to comet. Some comets even lack a visible dust tail! (Figure 13.12)

FIGURE 13.11   Tail develops on Comet NEAT (2004).

FIGURE 13.12   Gas tail of Comet Lovejoy (2011; left); dust tail of Comet McNaught (2007; right).

The dust tail may appear curved, sometimes so much so that from our Earthly point of view we can see it around the curve of the comet's orbit. Once the comet has rounded the Sun and is on its way back to the depths of space, the tail is in the lead. The tip of the dust tail looks like it is in front of the comet—a so-called anti-tail (Figure 13.13).

Larger dust particles remain stretched out along the comet's orbit, forming the comet trail. If the Earth passes through a trail, a meteor shower ensues. ➔ Chapter 9 ➔

FIGURE 13.13   Anti-tail of Comet Garradd (2012).

In 1997, astronomers observing Comet Hale-Bopp discovered a third type of tail consisting of neutral sodium atoms. This *sodium tail* is the straightest of all the tails, pushed away from the Sun by sunlight. The molecular source of the sodium atoms is still unknown.

Feeding all these tails is the nucleus that remains the core of the comet throughout its orbit. In 1950, American astronomer Fred Whipple ⟨1906–2004⟩ described them as 'dirty snowballs' because they are a roughly 50%/50% mix of water and other ices with gravel of rocks and dust. Comets have too little mass to be differentiated. Astronomers have identified all sorts of different ices in nuclei, including some organic molecules, such as the amino acid glycine (Figure 13.14).

Astronomers now consider a comet nucleus to be more of a rubble pile model than a snowball. The nucleus is not monolithic; rather, it consists of numerous smaller pieces that coalesced to form a larger body held together loosely by gravity. Large cavities could exist between the pieces in the interior, resulting in average densities as low as 0.3 g/cm^3. This structure would facilitate fragmentation events seen in comets, too.

Once a comet has moved out of the inner Planetary System, its tail disperses, its coma disappears, and it returns to its normal state, a mere solid nucleus. Gas and dust have left the comet, never to return. Comets lose material every time they revolve about the Sun. Eventually the comet nucleus is devoid of ices or its surface becomes layered with inert dust that insulates it and prevents sublimation. Astronomers have observed comets become fainter and fainter with each lap of the Sun. Some fade out like a bar of soap dissolving on the shower floor. Others first break into pieces like a snowball crumbling. The long-period comets, which have not made many passes by the Sun, retain most of their ice and generally give the most spectacular shows in our sky.

Halley's is a beat-up, short-period comet. It was barely visible to the unaided eye during its last visit to our planetary-system neighborhood, mostly due to an unfavorable viewing geometry, but had sentimental value, if nothing else. Still, we knew it was coming and prepared a flotilla of six space probes to meet it. The Japanese sent two probes, Sakigake and Suisei,[8] and the US managed to divert an existing solar probe to Halley's vicinity. The Soviet/French Vega 1 and 2 missions and the European Giotto[9] were the first to image the nucleus of a comet directly. Halley's Comet is oddly peanut shaped, and its surface is surprisingly dark with an albedo of 0.04! (Figure 13.15)

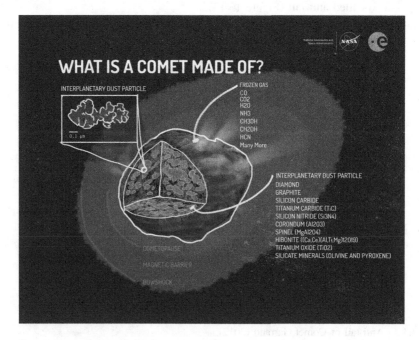

FIGURE 13.14 Typical composition of a comet nucleus.

FIGURE 13.15   Dust 'jets' emanating from the nucleus of Comet Halley (1986).

The Giotto pictures showed 'jets' of dust spewing forth from active spots on the sunlit side of Halley's surface. These jets explain why the comet often does not return exactly when predicted. In this case, another body is not changing the comet's orbit, but the comet itself effects its own orbit. The jets of escaping material act as rockets, pushing the comet in the opposite direction.

Halley's Comet brightened somewhat on its trip outward from the Sun. At this time, it should be dimming continuously. Perhaps its nucleus had a last gasp of activity when a subsurface pocket of gas bursts forth. Alternatively, maybe something hit the nucleus and ejected fresh material, enhancing the coma and its brightness. Astronomers will not know until they examine the comet when it comes our way again. Then, they will see, or not see, a new crater on its nuclear surface—in 2061. This behavior is not unique to Comet Halley. In 2007, Comet Holmes displayed an enormous outburst, outbound in the vicinity of the Asteroid Belt, when its brightness increased 1/2 millionfold! (Figure 13.16)

FIGURE 13.16   Comet Holmes (2007) in outburst.

Additional space probes have had close encounters with five comets. Results from these missions, combined with advanced space- and ground-based telescopes, have ushered in a golden age of cometary science.

NASA's Deep Space 1 mission flew by Comet Borrelly in 2001 and sent back even better images showing hills, steep slopes, cliffs, pits, mesas, and craters on a two-lobed nucleus. It was the first interplanetary craft to use an ion engine, opening a new era of space exploration.

In 2004, NASA's Stardust[10] mission flew through the coma of Comet Wild 2 gathering particles, which it returned to the Earth for examination. Some dust particles contained hydrated minerals, those that incorporate water. These are common on the Earth; an example is olivine. These minerals form at high temperatures so they must have forged near the Sun. That comets contain both 'ice and fire' materials was a major finding. Evidently, Solar System matter that formed near the Sun in its early history was transported outward and incorporated into comet nuclei (Figure 13.17).

Because this comet stuff has not changed much since the beginning of the Solar System, a sample could tell us what solar-system bodies were like before any geological activity took place. However, the fresh material lies below the comet's surface beyond remote observation. To this end, NASA's Deep Impact[11] mission fired a 100-kilogram (0.1 ton) copper projectile into a comet nucleus, hurling subsurface material outward so Earth-bound observers could study it (Figure 13.18).

ESA's Rosetta[12] mission rendezvoused and followed Comet Churyumov–Gerasimenko from 2014 until 2016 as it approached and then receded from the Sun. Rosetta carried a lander named Philae[13] that touched down on the surface of Churyumov–Gerasimenko, detecting more than a dozen organic compounds, including the amino acid glycine. The orbiter mapped the gnarled surface in unprecedented detail with some images taken mere kilometers from the comet. In particular, the Imhotep[14] region is remarkably diverse with fine-grained plains, boulders, rocky terrains, basins, icy patches, and fractured terraces.

Together, Rosetta's lander and orbiter revealed that the two-lobed nucleus was likely formed from a collision of two smaller bodies early in its history, a contact binary. At mission's end, the orbiter was guided down to the comet's surface, gathering data until impact. The wealth of information collected by Rosetta will ensure scientific treasures for years to come (Figure 13.19).

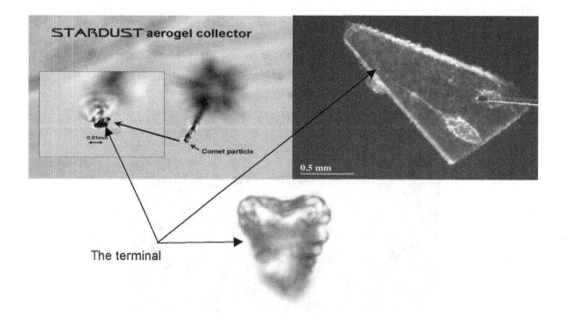

FIGURE 13.17 Heart-shaped comet particle collected by Stardust.

FIGURE 13.18   Deep Impact sends an impactor crashing into Comet Tempel 1 (2005).

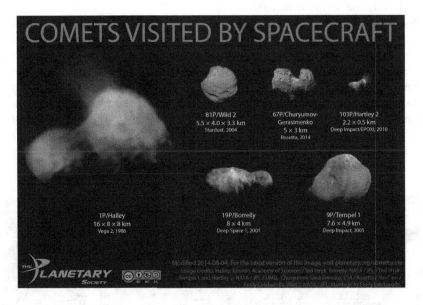

FIGURE 13.19   Comets visited by space probes: Nuclei are in fact cratered.

ESA's Solar and Heliospheric Observatory [SOHO] watches the Sun and its nearby environment. It has discovered well over *4,000* sun-grazers—comets that travel up to 70 times closer to the Sun than the planet Mercury. SOHO's images even show dozens of comets plummeting into the fiery Sun each year. Sometimes these 'suicidal' comets come in groups. Were they once part of a single, larger comet? Current thought points to a parent Great Comet as the origin of many fragments, possibly seen by Aristotle in 371 BCE.

As we have discussed throughout, collisions are a fact of life in the Planetary System. Collisions among planetesimals are interesting, but, just as we prefer asteroids to keep their distance, we do not want a comet too near *us*. A comet impact would be as devastating as an impact from a similar mass asteroid.

In the mid-1990s, a comet named Shoemaker-Levy-9 [SL-9] was tidally pulled apart when it fell under the influence of Jupiter's immense gravity; 'fluffy' comets have little tensile strength and are easily disrupted. The orbit of the resulting pieces was so distorted by Jupiter that they each *struck* the planet in turn, each with the energy of many nuclear bombs. Every collision left a dark spot ('bruise') in Jupiter's atmosphere, but only for a while. The comet, which was, of course, completely destroyed, got the worst end of this encounter. The crash was ideal because it allowed scientists to study violent impacts—each as large as the putative dino-killing impact of 66 million years ago—but on somebody else's planet (Figures 13.20 and 13.21).

Thankfully, we (on the Earth) have a massive guardian nearby (Jupiter) to take some of the hits for us. But before you pat Jupiter on the back as a hero, it also deflects some comets to the inner Planetary System, increasing the odds that they may collide with us!

As we leave these fascinating objects, let us point out the need for consensus definitions of a comet (phenomenological) and comet nucleus (physical). None are universally accepted and English-language dictionaries are woefully lacking from a scientific perspective. We now have a very sophisticated and diverse inventory of small bodies that challenge our previous notions of comets including, Main Belt Comets/Active Asteroids, Centaurs, Jupiter's Trojan asteroids (co-mets?), Kuiper Belt Objects (e.g., Arrokoth), interstellar comets (e.g., Borisov), sun-grazers, and others. We also have detailed information of several comet nuclei from spacecraft encounters. As new information is gained, we must revise our understanding of comets as part of our activities as scientists following the scientific method.

Currently, comets are classified as objects that show some sort of activity at their time of discovery, usually dust (not gas), and have orbital motions with respect to the Sun (except in-terstellar comets!). This is much like the ancients did in their time. Individual researchers or the Minor Planet Center at Smithsonian Astrophysical Observatory make the final determination. This situation needs to be revisited and is long overdue. A new definition needs to take into account physical and chemical properties, dynamics, cosmogony (time and place of origin), activity me-chanisms, and other properties, in addition to observational aspects. The relationship of comets to asteroids is also part of this discussion since evidence points to a continuum of bodies between these objects, rather than two separate and distinct groups. If you would like to contribute your opinions to this discussion, please let the authors know.[15]

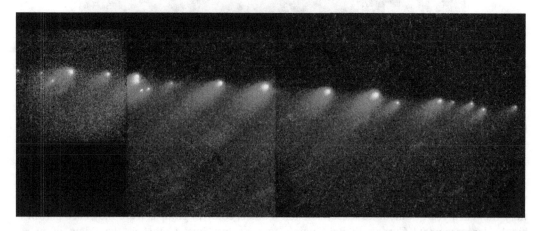

FIGURE 13.20 Comet Shoemaker-Levy 9 [SL-9] near Jupiter, after its disassociation by the giant planet's gravity (1994).

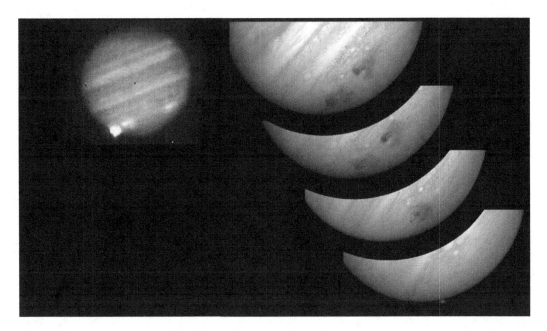

FIGURE 13.21   Infrared image showing a fragment of SL-9 explosively striking Jupiter (left); evolution of SL-9 impact spots on Jupiter (right).

NOTES

1 A personal quirk of ours: It is not 'Haley's Comet.' Bill Haley and His Comets was a popular rock-and-roll band of the 1950s. But the pronunciation is different; Halley is a short 'a.' Evidence suggests that the man himself pronounced it more like 'hall[w]ay.'

2 The orbital period has varied from 76 to 79 years due to planetary gravitational influences.

3 Half-human, half-horse creatures of Greek myths.

4 Wisest of centaurs, who trained many Greek heroes.

5 Wife of the centaur Chiron.

6 Hawai'ian for 'messenger from afar arriving first.'

7 Shared the Nobel Prize in Physics in 1970 for fundamental work about the relationships among electrons, ions, electricity, and magnetism.

8 In Japanese, sakigake means pathfinder; suisei means comet.

9 In one of Giotto di Bondone's ⟨circa 1267–1337⟩ works, he painted a nativity scene depicting the Star of Bethlehem as a comet, possibly inspired by Halley's Comet, which was visible during his lifetime.

10 Stardust was redirected to Comet Tempel 1 in 2011 before its fuel was expended.

11 Deep Impact likewise was 'recycled,' encountering Comet Hartley 2 in 2010.

12 From the Rosetta Stone, used to decipher Egyptian hieroglyphs.

13 Named after the Philae obelisk, bearing an inscription like the Rosetta Stone.

14 Meaning 'the one who comes in peace,' an Egyptian chancellor to the pharaoh in late 27th century BCE.

15 A lively discussion is taking place about specific definitions of a 'comet' and 'comet nucleus' and their relationship to asteroids. If you would like to contribute your ideas, a brief survey can be found in the Further Readings appendix for both professionals and laypeople alike.

14 From Kuiper Belt to Oort Cloud

Astronomers know what becomes of a comet that spends too much time near the Sun. They also know that many comets reside within the Planetary System, having orbits that do not extend pass that of Neptune. Aside from a few interstellar ringers, where do comets come from in the first place? What is their origin? They must have formed outside the snow line, the distance from the Sun at which ice remains frozen, currently at the distance of Jupiter's orbit (Figure 14.1).

Potential comet nuclei also orbit beyond Neptune in a region known as the Kuiper Belt, named after a pioneer planetary scientist of the 1940s and 1950s, Dutch-American Gerard Kuiper ⟨1905–1973⟩ . He proposed it in 1951 after asking a simple question: Why should the Solar System end at Pluto? He and others could find no reason why it should but neither did they know what might reside there. Economist and engineer Kenneth Edgeworth ⟨1880–1972⟩ had a similar idea before Kuiper in 1943, so some astronomers prefer the name Edgeworth-Kuiper Belt.[1] These vague, early notions were hard to test observationally.

Not until 1980 when Uruguayan astronomer Julio Fernández argued more convincingly that short-period comets might come from a disk-shaped region beyond Neptune, did the Kuiper Belt take shape. Still, more than a decade passed before someone actually looked carefully for evidence of these icy bodies (Figure 14.2).

No longer hypothetical, the Kuiper Belt is ring of small rocky and icy bodies, including Pluto and at least three other dwarf planets, Eris, Haumea, and Makemake. It begins at 30 au and extends to several hundred au from the Sun in the ecliptic plane. Its densest part lies between 40 and 48 au. Rather than being flat, the Kuiper Belt is inflated like the inner tube of a tire.

The first Kuiper Belt Object [KBO] photographed was Pluto, discovered in 1930, but its role in this region was not understood then. Beginning in 1992, large Earth-based telescopes finally detected more of largest members of the Kuiper Belt. British-American David Jewitt and Asian-American Jane Luu found the next KBO on 30 August 1992 after searching for 6 years. Slowly the floodgates opened and now more than 1,000 KBOs are known. Astronomers estimate that more than 100,000 bodies larger than 100 kilometers (60 miles) across exist in this icy realm. Even with our largest telescopes, only the biggest and closest KBOs can be seen directly (Figure 14.3).

Because astronomers expect large bodies to be rare compared to smaller ones, the Kuiper Belt must be a repository for an even greater number of comet nuclei. If so, there are roughly 1,000 KBOs for every asteroid of similar size, filling a region about 1,000 times the volume of the Asteroid Belt. Indeed, telescopic survey images (e.g., HST) suggest that hundreds of thousands more KBOs to be detected, while hundreds of millions more remain invisible. The mass of the Kuiper Belt may exceed that of the Asteroid Belt by hundreds of times. However, its total mass is only a modest 10% that of the Earth. The Kuiper Belt may have been much more massive, about 20 to 30 Earth masses, but most of that has been lost over the age of the Solar System.

Collisions or close encounters with other KBOs deflect some objects out of the Kuiper Belt, either ejected to interstellar space or hurtling into the inner Planetary System. As these deflected bodies approach the Sun, their volatile ices will sublimate to form comets with their familiar comae and tails.

The orbits of KBOs fall into several groups. About half of the known bodies have almost circular orbits beyond Neptune, called Classical KBOs. Many of these have stable resonances with Neptune such that their orbital periods are simple multiples of Neptune's. For example, Pluto falls into this group since it makes two orbits for every three orbits made by Neptune. It, along with thousands of other KBOs, are in a 3:2 resonance with Neptune; these are the Plutinos. Many other

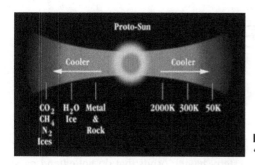

FIGURE 14.1 Diagram showing the Solar System 'snow line.'

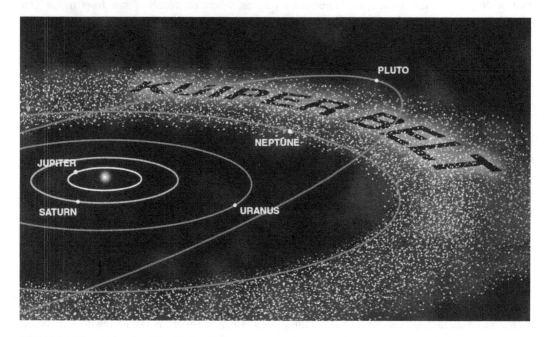

FIGURE 14.2 Diagram of the Kuiper Belt.

resonances are also seen, such as 1:1 (Neptune's Trojans), 2:1, 4:3, *etc*. As predicted by the Nice Model, Neptune forced these bodies into resonances as it migrated slowly outwards from its formation at about 15–20 au to its present distance of 30 au. ➤ See Chapter 1 ➤

During the elaborate kabuki dance of the giant planets predicted by the Nice model, Uranus and Neptune shepherded some of the Kuiper Belt to its present location and may have even traded places with each other. In doing, they also formed the Scattered Disk, a part of the Kuiper Belt that extends beyond 100 au. It contains objects with very elliptical orbits not in resonance with Neptune and at higher inclinations. Eris is its largest known object. Astronomers think the Scattered Disk is the principal source of short-period comets and the Centaurs.

We now recognize that the Solar System is much more dynamic and chaotic than the previously held clockwork notion, where bodies moved along predictable orbits for billions of years. Consequently, tracing the history of the Solar System is difficult. During an outward planetary migration as suggested by the Nice Model, the architecture of the Solar System would be catastrophically upset if two planets fell into a resonance. Such an event would disrupt the early massive Kuiper Belt, showering the inner Planetary System with a swarm of comets and creating giant impacts on planetary surfaces, such as those seen on the Moon and Mercury. The remnant Kuiper Belt would have lost most of its mass and appear as it does today.

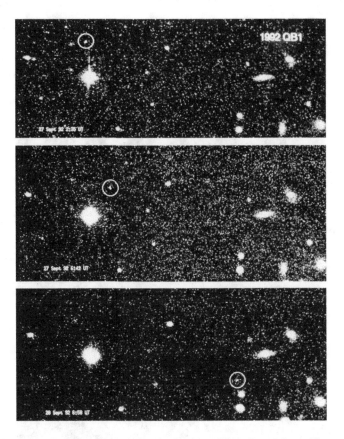

FIGURE 14.3 Discovery of a Kuiper Belt Object by Jewitt and Luu. Its motion with respect to background stars gives it away.

NASA launched the New Horizons space probe in 2006 to study Pluto and its moons and to reconnoiter the Kuiper Belt generally. This first mission to the icy bodies at the outer limits of our Planetary System is revealing their composition, structure, and atmospheres in greater detail than could be achieved with ground-based instruments.

Reaching the plutonian system in 2015, New Horizons spent 6 months observing the largest known KBO and its satellites. Pluto is also a dwarf planet along with fellow KBOs Eris, Haumea, and Makemake.[2] Astronomers think other bodies are similarly large, but, even in powerful telescopes, they appear to be mere points from the Earth. So far, the IAU has not confirmed any other dwarf planets (Figure 14.4).

For the better part of a century, people considered Pluto one of the major planets. One-time amateur astronomer, Clyde Tombaugh ⟨1906–1997⟩ discovered it in 1930. Lowell Observatory had hired him to seek a ninth planet in the Solar System. He did so by using astronomical photography: comparing images made on two different dates and finding the point of light that had changed positions with respect to the constant stars (Figure 14.5).

After its discovery, learning anything more about Pluto was difficult. At an average distance of 40 au, it is far away and takes 248 years to complete an orbit. With a diameter only a little more than 2,400 kilometers (1,500 miles), it is smaller than several satellites, including our Moon. In 1978, the discovery of its first satellite, Charon,[3] enabled the measurement Pluto's mass and density. The answer is intermediate: not the 1 g/cm^3 of the other outer solar-system planets, but not the 4 g/cm^3 of an inner solar-system body either. It was about 2 g/cm^3, like the Icy Satellites of the giant planets, which suggests it is made of rock and ice. Pluto's ice makes it easier to find; it is

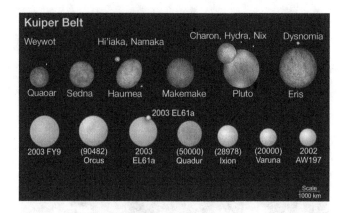

FIGURE 14.4   Illustration of some known Kuiper Belt Objects (to scale).

FIGURE 14.5   Clyde Tombaugh at Lowell Observatory *circa* 1930.

shiny with an albedo of 0.72. Pluto is probably an ice ball with a core of rock and metal, a distasteful 'Tootsie Pop' (Figures 14.6 and 14.7).

Even at Pluto's low temperature, which averages −225°C (−373°F), it has a thin atmosphere. Because Pluto's gravity should be too low to keep hold of gases, its atmosphere cannot have been there very long. Yet, if a world is mostly ice that constantly sublimates, eventually we would have no Pluto! However, Pluto's orbit is quite eccentric, especially compared to the major planets. It is warmer at perihelion than at aphelion 124 years later. Ice turned to gas at perihelion may be refrozen at aphelion and then fall back onto the surface before it can escape. Thus, Pluto has a *temporary* atmosphere at perihelion.

The composition of this tenuous atmosphere is mainly nitrogen, with small amounts of hydrocarbons (e.g., methane, acetylene, ethylene, and ethane). Layers of haze cool the atmosphere.

By the way, the eccentricity of Pluto's orbit is so great that, as it approaches the Sun, it crosses the orbit of Neptune. Fear no collision, though; Pluto's orbit is also more inclined to the ecliptic than the those of the planets. In fact, because Pluto has a 3:2 orbital resonance with Neptune, when Pluto is at perihelion, Neptune is on the opposite side of the Sun. Pluto actually approaches Uranus closer than it does Neptune!

If Pluto is not the smallest planet, might it be the largest comet? While it does sublimate a thin atmosphere, it never gets close enough to the Sun to form long ion and dust tails. Moreover, Pluto is too massive to allow gas and dust to escape its gravitational pull freely as is characteristic of comets. Pluto also has the uncometary characteristic of a large natural satellite, Charon, and four smaller ones (Hydra,[4] Nix,[5] Styx,[6] and Kerberos[7]).

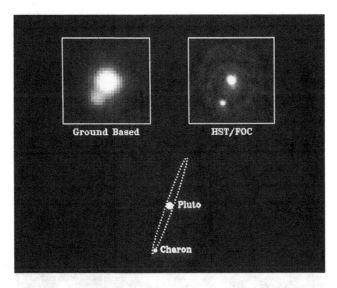

FIGURE 14.6   First appearing as a 'mountain' on Pluto, Charon later was resolved as a satellite. Its orbital period allowed determination of the combined mass of the Pluto/Charon system.

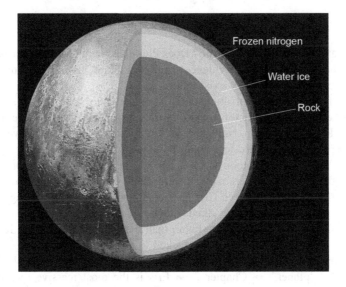

FIGURE 14.7   Cross section of Pluto.

As the first mission to Pluto, the New Horizons encounter greatly expanded our knowledge. The returned images transformed Pluto from an astronomical object to a geological one, a point of light to a new world. The probe revealed surface details that can never be detected from Earth. Crater counts on its surface show that it has been geologically active throughout the past 4 billion years. Sputnik Planita, a massive nitrogen-methane ice basin larger than Texas, is devoid of any craters and estimated to be very young—no more than 10 million years old. It is the largest glacier in the Solar System. Evidence exists of even younger features from recent cryovolcanism. Pluto's bladed terrain is littered with skyscrapers of methane ice, gigantic shards reaching heights of 500 meters (1,600 feet). The variations of surface composition on Pluto—from nitrogen-rich, to methane-rich, to water-rich areas—are not seen elsewhere in the outer Planetary System (Figure 14.8).

FIGURE 14.8   Pluto imaged by the New Horizons space probe.

What lies beneath the icy surface of Pluto? Does Pluto possess a global ocean as do a number of Icy Satellites of the Jovian Planets? Evidence from New Horizons indicates that it likely had a global ocean in the past that still may exist today.

Observations of Pluto's nitrogen atmosphere show organic haze layers similar to Titan's. Unexpectedly, haze is the major factor governing the atmospheric temperature, unlike any other atmosphere in the Planetary System. New Horizons found that interactions between the solar wind and Pluto's atmosphere are confined to a small region on the dayside.

In contrast to Pluto, its large moon Charon has an ancient surface. Smooth, large plains on Charon are likely the result of vast cryovolcanic eruptions, thought to be related to the freezing of an internal ocean that globally ruptured Charon's crust about 4 billion years ago.

Pluto's smaller satellites have higher albedos than typical KBOs, providing evidence that they may have formed from a disk around Pluto rather than being captured from the Kuiper Belt. Their rotation rates vary greatly and are not tidally locked to Pluto, like Charon.

The discovery of Eris in 2003[8] challenged Pluto's original designation as a planet and led to the modern definition of 'planet.' ➤ Chapter 2 ➤ Eris is the most massive dwarf planet, though slightly smaller in diameter than Pluto. It orbits the Sun every 559 years on a highly eccentric orbit at an average distance of 68 au. Because its orbit nearly crosses that of Neptune, Eris also may grow a temporary atmosphere like Pluto. The surface of Eris is bright with an albedo of 0.55. It has at least one moon, Dysnomia.[9]

Next to be discovered was Haumea in 2004. It orbits the Sun at an average distance of 43 au. Its spins rapidly, once every four hours, which is the fastest known for large KBOs. Such a fast rotation produces a unique ellipsoidal shape; Haumea's equatorial bulge is twice as large as the distance between its poles. This dwarf planet has a red marking on its surface. With Pluto, it has more than more than one moon: Hi'iaka[10] and Namaka.[11]

In 2005, Makemake was the last of the dwarf planets discovered so far; it orbits the Sun at an average distance of 46 au. Like Haumea, its orbit is highly inclined to the ecliptic. Astronomers determined the mass of Makemake using the orbit of a provisional moon. It is about 2/3 the size and 1/3 the mass of Pluto.

Other large KBOs include Quaoar,[12] Salacia,[13] Orcus,[14] and Gonggong[15] (nicknamed Snow White because of its high albedo). All these have peculiarly inclined, elliptical orbits typical of the smaller KBOs. The densities of large KBOs range from 1.5 (Orcus) to 2.5 g/cm^3 (Eris), certainly qualifying them as icy bodies. Each of these also has at least one moon. With more detailed observations, the IAU likely will classify all these KBOs as dwarf planets. Some day we may refer to Snow White and the seven dwarfs!

As fascinating as the dwarf planets and potential dwarf planets are, the Kuiper Belt contains many more small, irregularly shaped icy bodies. Having flown by Pluto, NASA redirected the New Horizons space probe to another member of the Kuiper Belt. On 1 January 2019, New Horizons flew by Arrokoth,[16] the first small body encountered in the Kuiper Belt, passing at a distance of 3,500 kilometers (2,300 miles). New Horizons captured many scientific treasures. Most notably, it returned the first images of a primordial comet nucleus, one that formed where it is currently located and has never had a near encounter with our fiery star. Preserved in this pristine state, Arrokoth provides clues about our early history, some 4.6 billion years ago (Figure 14.9).

This sensational mission revealed a two-lobed body joined by a small 'neck'; it is a contact binary similar to Comet Churyumov–Gerasimenko. This shape likely resulted from a very slow collision, where two separate bodies stuck. Arrokoth's size is about 34 kilometers (21 miles) in the longest dimension, slightly larger than the average comet nucleus but certainly smaller than Hale–Bopp, with a spin period of 16 hours or so.

Arrokoth shows surface properties consistent with comet nuclei, such as very dark albedo with some minor variations and a bright ring at the neck—a mottled and ruddy terrain. Similarly, it has hills, slopes, plateaus, and craters, but with less relief due to its unprocessed nature. Arrokoth's surface is slightly reddish from irradiation of ices during its long storage stint, similar to many other KBOs. Although its detailed composition is still undetermined, mission scientists have identified ices of water, methanol, and organic molecules mixed with rocks on its surface.

At Arrokoth's distance of 43 au from the Sun, surface temperatures are frigid, about −230°C (−380°F), but extremely volatile ices of carbon monoxide, nitrogen, oxygen, and methane have sublimated from its surface. New Horizons is still transmitting its data, so stay tuned for more revelations about this unusual body and perhaps a cousin, since NASA may redirect the spacecraft to yet another KBO!

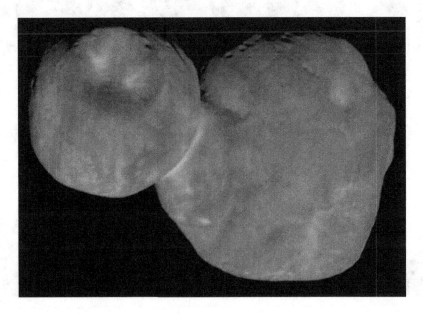

FIGURE 14.9 Arrokoth imaged by New Horizons.

While a small KBO, like Arrokoth, may be a future short-period comet, long-period comets cannot come from the relatively nearby Kuiper Belt. Their highly eccentric orbits with periods of up to millions of years stretch far beyond that region of our Solar System. With orbits also randomly inclined, the long-period comets must originate in a remote spherical source at the extreme limits of the Sun's gravitational influence. From where do they come?

Dutch astronomer Jan Oort 〈1900–1992〉 first postulated such an outermost region of the Solar System in 1950. He noticed that almost all long-period comets go out to at least 50,000 au. To him, this distance seemed more than a coincidence. Oort envisioned a vast reservoir, a shell of dirty snowballs surrounding the Solar System at that distance. The shell may extend out to 150,000 au in radius, the limit of the Sun's gravitational influence where other stars compete for control. Within this shell, potential comet nuclei swarm. We cannot see the Oort Cloud. It is too far away. But it must be there.

Here we make the important distinction between the Solar System and the Planetary System. Oort showed that the influence of the Sun extends to at least 50,000 au, far beyond the realm of planets (about 30 au). When NASA claims that a certain space probe has left the Solar System, they actually mean that it has left the Planetary System at 30 au, or at best the influence of the solar wind at about 120 au. Please remind them that it has quite a way to go before leaving the Solar System! (Figure 14.10)

Today, astronomers divide the Oort Cloud into two regions. The doughnut-shaped Inner Cloud extends from the inner edge at about 2,000 au from the Sun out to about 20,000 au. It smoothly connects the Kuiper Belt to the surrounding Outer Cloud. The Outer Cloud is the spherical shell envisioned by Oort; the outer edge of this region is tenuous. Astronomers associate several thousand known long-period comets with Outer Cloud origins. The Inner Cloud is probably the source of the Halley-type comets, with periods between 20 and 200 years, insofar as they require a closer origin than the long-period comets. Sedna,[17] a potential dwarf planet, is also a possible member of the Inner Cloud.

By far, most of these bodies—more than a trillion of them!—stay in the Oort Cloud. They slowly move around the Sun, always at a vast distance, in nearly circular orbits. They are as cold as

FIGURE 14.10 Diagram of the Oort Cloud.

space gets: about −270°C (−450°F). Occasionally, however, something disturbs these loosely bound objects from their hibernation in the Oort Cloud and sends them careening into the Planetary System.

The gravitational influence of a wandering interstellar cloud of gas or the shockwave from a nearby exploding star, a supernova, could disrupt a portion of the Oort Cloud. Another possibility is the gravitational influence of a traveling nearby star. Once every 3–10 million years, a star sweeps within 100,000 au of the Sun, which is close enough to produce a noticeable effect. The star Gliese 720 actually will pass through the Oort Cloud—in a million years (Figure 14.11).

Our Sun, along with hundreds of billions of other stars, slowly revolves around the center of a huge disk of stars called the Milky Way Galaxy. Every time the Sun passes through a dense part of the disk, the disk's gravity may nudge the Oort Cloud (Figure 14.12).

Any of these mechanisms could supply us with enough comets for 10,000 years. Regardless, nuclei are bumped out of the cloud. Some escape the Solar System entirely. It even may be possible for such a comet to travel from the influence of one star to another, similar to Oumuamua and Borisov. ➤ Chapter 13 ➤ Most likely, however, the nucleus plunges inward becoming a long-period comet.

Where did the Oort Cloud comets originate? These objects could not have formed in their current position, because the material in the proto-solar nebula at these large distances would have been too sparse to aggregate. Rather, astronomers think they originally formed out of the denser gas and dust present in the inner nebula, coalescing into nuclei as the leftovers from the formation of our Solar System.

These leftovers were subjected to the gravitational pull of the now-massive planets, and some were sent to the inner Planetary System where they vanished long ago. Recall that the Nice Model posits that the Giant Planets originated much closer to the Sun. As they migrated outward, a flood of comets was unleashed throughout the Solar System as part of the Heavy Bombardment Era. ➤ Chapters 1 and 4 ➤ Many may have collided with the early Earth, supplying our planet with water and organic compounds. The outer giants would have also consumed nearby comets, which contributed to their growth. Many comets escaped from the Solar System entirely, becoming

FIGURE 14.11 Shockwave from a supernova.

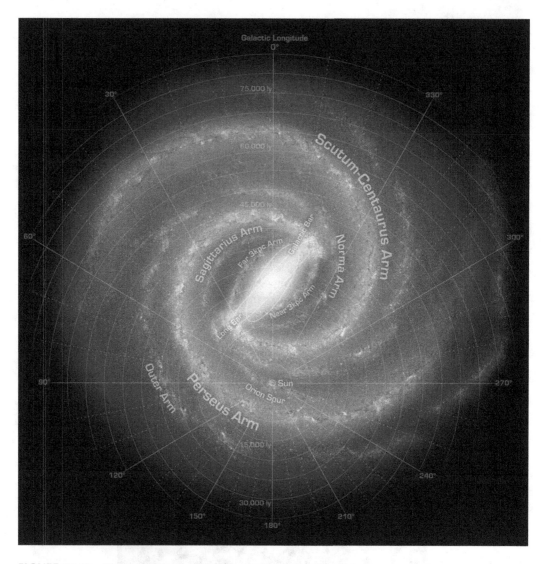

FIGURE 14.12   Diagram of the Sun's position in the Milky Way Galaxy.

interstellar comets bound for parts unknown in our Galaxy. Finally, some remained weakly bound to the Sun at large distances in the Oort Cloud.

We have now taken you to the outer edge of our Solar System, where the Sun's gravitational pull can no longer bind matter. Now let us instead journey to the center of our Solar System and look at our personal star, the Sun (Figure 14.13).

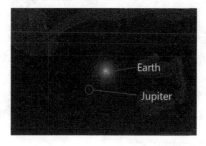

FIGURE 14.13   Voyager 2 space probe 'looks' back at the Sun from a distance of 122 au.

NOTES

1 Others simply call it the Trans-Neptunian Region, bodies therein being Trans-Neptunian Objects [TNOs].
2 The dwarf planets of the Kuiper Belt are collectively called Plutoids as a consolation to Pluto after losing its planethood.
3 In Greek mythology, the ferryman who transports souls across the rivers of the underworld to the realm of Hades.
4 In Greek mythology, a monster with the body of a serpent and nine heads that guards the realm of Hades.
5 In Greek mythology, Nyx was the goddess of night, who resides in the realm of Hades. However, to avoid confusion with the previously named asteroid Nyx, the IAU chose an alternative spelling.
6 In Greek mythology, a river that bounds the realm of Hades.
7 In Greek mythology, a dog with three heads that guarded the realm of Hades.
8 The discovery team delayed their official announcement until 2005.
9 The Greek goddess of lawlessness.
10 Patron goddess of the Island of Hawai'i.
11 Water spirit in Hawaiian mythology.
12 Tongva creation god.
13 Roman goddess of salt water.
14 Etruscan god of the underworld.
15 Chinese water god.
16 Powhatan/Algonquian word meaning 'sky.'
17 Inuit sea goddess.

15 Sol, Our Sun

When we call it the Solar System, we are naming it after its grandest member: the Sun. 'Sol' is the Roman god that personified the Sun. It is the brightest object in the sky and the hottest body in the Solar System. Because of its dominance in the sky and its life-giving heat and light, many ancient cultures considered it a deity to be worshipped. The Romans sacrificed to Sol on August 9th for a good harvest (Figure 15.1).

We usually think of the Sun as a very reliable partner, rising as it does every morning and setting every evening and supplying relatively constant energy to sustain all life on the Earth. Yet, inside and out, it is a surprisingly violent place, where destruction (and construction) take place on an immense and blindingly fast scale (Figure 15.2).

The Sun is entirely different from anything else that we have discussed in the Solar System. Still, it is, in fact, just another star among hundreds of billions in the Milky Way Galaxy alone: average in size and other stellar properties, and middle-aged! Its retinue of followers (i.e., planets, asteroids, comets, and others) is not at all unusual either. Astronomers now know of thousands of other stars that have their own attendant planetary systems. ☞ Chapter 16 ☞

The Sun is enormous, but it is also 1 au away from us. In the sky, the Sun appears to be about the same angular size as the Moon. Recall that this coincidence enables total solar eclipses when the Moon travels in front of the Sun. The Sun is 1.4 million kilometers (870,000 miles) in diameter. Ten Jupiters or about 109 Earths could stretch across its equator.

More impressive is a comparison of these bodies' volumes since around a thousand Jupiters (or 1.3 million Earths) could fit inside the Sun. More than 99.8% of all the matter in the Solar System resides in the Sun; it is 1,000 times more massive than the largest planet, Jupiter. Of course, this mass provides the powerful gravitational attraction that binds the Solar System to the Sun. If you could stand on the 'surface' of the Sun, its gravity would be 28 times that of Earth. Yet, its average density is only 1.4 g/cm^3, reflecting its gaseous nature (Figure 15.3).

The Sun is 73% hydrogen and 25% helium, with traces of the other elements making up the remaining. In fact, all 92 naturally occurring elements have been found in the Sun. In 1868, English astronomer Norman Lockyer ⟨1836–1920⟩ identified an element in the Sun that was unknown on Earth at the time; he called it helium after Helios. Compositionally, the Sun is a lot like Jupiter.

If Jupiter had grown bigger and more massive, we possibly could be living in a stellar system with two suns, like *Star Wars*'s Tatooine. Actually, Jupiter missed stardom by quite a bit, needing about 80 times its current mass to be a small red dwarf star.

Like the Gas Giants, the Sun rotates differentially, with a period of 25 days at the equator, decreasing to 35 days at the poles. Turning in the same direction as its planets, its average rotation rate is once per 27 days.

The similarities between the Sun and Jupiter end here, though. The difference between the Sun and any planet is overwhelming. Talk about worlds of fire; the Sun puts all others to shame! It is only the central Sun that is so hot as to turn all elements into gases. They are not the normal gases that we encounter daily on the Earth, rather they consist of a fourth state of matter called plasma, a mixture of electrically charged atoms and electrons that behaves very differently from normal gases. A plasma is so hot that it blazes with electromagnetic radiation; this solar radiance illuminates the rest of the Solar System (Figure 15.4).

Why is the Sun so hot? It has an effective surface temperature of 5,500°C (10,000°F), quite enough to radiate light and heat the equivalent of four *septillion*[1] 100-Watt light bulbs! What keeps

FIGURE 15.1   Surya, Javanese sun god.

it going? The Sun presumably is at least as old as the planets that formed to revolve about it. The oldest Earth rocks formed billions of years ago. Even though at least billions of years old, the Sun (thankfully) shows no sign of wearing out.

Hydrogen is flammable, but it only burns in the presence of oxygen, of which there are scant amounts available in the Sun. For a long while, astronomers thought that the Sun was gradually contracting, converting potential energy into heat. That would work for hundreds of thousands of years, but not billions.

The only option left is nuclear energy, as pointed out by the British astronomer Arthur Eddington ⟨1882–1944⟩ but embodied in the famous equation we have all heard mentioned but perhaps not contemplated (Figure 15.5).

$E = mc^2$ was the work of Einstein, a consequence of his Theory of Special Relativity. The most important part of this equation is not the E, which stands for energy. It is not the m, which stands for mass. It is not even c, a large constant representing the speed of light. No, the most important thing in Einstein's equation is the equals sign. We think of mass as something tangible—all the objects in our Universe (Figure 15.6).

We think of energy as something more ephemeral: light, heat, or motion. The equation eliminates this somewhat arbitrary distinction. Under the right conditions, energy turns into mass, and mass can be turned into energy.

FIGURE 15.2   The Sun has been rising and setting in our sky for over 4 billion years.

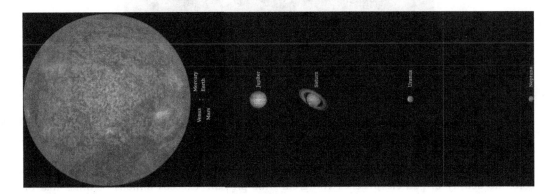

FIGURE 15.3   Comparison of sizes of the Sun and planets (to scale).

If you bring a kilogram of refuse to the trash pile in your alleyway and light it, chemical fire will transform the contents into ash, smoke, gas, and charcoal. However, if gathered and weighed, together, the mass of this residue still will remain 1 kilogram. Not so in a nuclear reaction; the fuel is annihilated for good (Figure 15.7).

The constant c is a big number. Its square is a *really* big number. Our equation thus tells us that it takes little mass to produce prodigious energy. The Sun has plenty of mass, and its nuclear

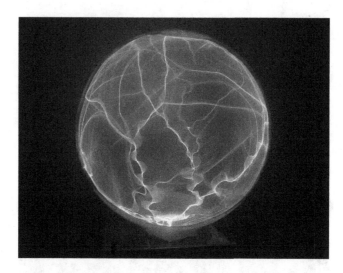

FIGURE 15.4   Trapped plasma used as a work of art.

FIGURE 15.5 $E = mc^2$.

engine will continue to run for billions of years more. In fact, it has just hit its stride, having increased in luminosity by 30% since its formation.

Within a solar-mass gas, in hydrostatic equilibrium, conditions for nuclear reactions are met in the Sun's so-called *core*, roughly the inner 25% of the Sun's diameter. (We use the term 'core' here in a slightly different sense than we have before; the Sun's core contains 30% of its mass.) As you hypothetically descend into the Sun, its temperature, pressure, and density all increase. In the core, the density of particles reaches 160 g/cm^3, the pressure exceeds billions of times that of Earth's atmosphere, and the temperature rises to 15 million °C (27 million °F). Such conditions cause subatomic particles to fuse together, losing mass in the process. This missing mass has not simply been changed into another form. It is truly gone, converted into energy.

To understand this process, we need to 'talk' about nuclear physics for a while. First, there are four fundamental forces in nature. We have discussed gravity and mentioned the electromagnetic force, both very long-range forces. Now, we introduce the remaining two forces, which operate over distances comparable with the nucleus of an atom, the strong and weak nuclear forces. In fact, 'nuclear' refers to the nucleus of an atom; the solar core is so hot that the electrons normally bound to the nucleus have been stripped away.

FIGURE 15.6   Albert Einstein.

Second, this energy-conversion process is called nuclear fusion, the joining of light elements to make heavier ones. Fusion is the inverse of nuclear fission where a large atom, like uranium or plutonium, is split into smaller atomic fragments.

As mentioned earlier, the Sun is primarily composed of hydrogen in its plasma form, that is, a proton [^{1}H] and an electron [e^-]. Because solar gravity compresses this hot, dense sea of protons and electrons, collisions between them are very frequent. Protons and electrons naturally attract because they carry opposite electric charges, but, in the extremely hot solar core, they deflect instead of bonding. Two protons on a collision course naturally repel since they both have positive charges—to a point. When the temperature is high enough, starting at around 10 million °C (18 million °F), the protons overcome their electrostatic repulsion and come close enough for a very short-range force, the strong nuclear force, to bind or fuse them together. Hence, fusion occurs.

You can think of the strong nuclear force as the glue that sticks protons together. It keeps atomic nuclei, consisting of protons and neutrons in close proximity, from falling apart.

Fusing two hydrogen nuclei, or protons, is just the first nuclear reaction that initiates the so-called proton-proton [p-p] cycle, the series of reactions that converts hydrogen to helium, resulting in a tremendous release of energy.

Let us go through the complete p-p process. During the fusion of the two protons, one converts into a neutron. Two exotic byproduct particles are released: a neutrino [v_e] and a positron [e^+]. A neutrino is a neutral and nearly-massless particle that travels at almost the speed of light. A

FIGURE 15.7   Chemical burning results in a change in the nature of matter, but it does not convert matter to energy.

positron behaves like a positively-charged electron; it is called the antiparticle of the electron. Antimatter is real and is being produced in the Sun's core in large quantities. Unlike the ghostly neutrino that escapes directly to the 'surface,' the positron cannot survive in the crowded core. Antiparticles annihilate when they encounter their counterparts (in this case an electron) in a blinding flash of energy, resulting a two gamma-ray[2] photons [γ]. A photon is a unit of electro-magnetic radiation or light.

The proton-neutron pair produced in the p-p cycle is called deuterium [^{2}D], a heavy isotope of hydrogen. Atomic isotopes are identical elements, except for varying number of neutrons in their nuclei. Summarizing as an equation, solar physicists write

$$^1H + {}^1H \rightarrow {}^2D + e^+ + \nu_e$$

This initial step is a bottleneck in the p-p cycle. Protons naturally decay into neutrons by influence of the weak nuclear force. However, doing so takes a very long time, up to a billion years for a single decay. Even so, so many collisions occur in the core every second that a prodigious amount of loose deuterium still accumulates.

The production of a free positron is followed immediately by electron-positron annihilation:

$$e^- + e^+ \rightarrow 2\gamma.$$

Next, a third proton fuses with the deuterium to make a light form of helium [^{3}He], an isotope of regular helium [^{4}He]. Another gamma-ray photon is released.

$$^2D + {}^1H \rightarrow {}^3He + \gamma$$

In the last step, two light helium nuclei fuse into regular helium, containing two protons and two neutrons, and recycle two additional protons back into the core plasma, where they are available for the next round of p-p fusion.

$$^3He + {}^3He \rightarrow {}^4He + {}^1H + {}^1H$$

A net of four protons is converted into one helium nucleus containing two protons and two neutrons. Several other particles and photons are released in the process. If the four protons were put on a balance scale opposite the helium nucleus, it would tip towards the protons. The protons are more massive, so where did the missing mass go? It was converted into energy via Einstein's equation in the form of gamma rays and neutrinos.

In the Sun, the p-p cycle converts 620 billion kilograms (685 million tons) of hydrogen into helium every second, releasing an enormous amount of energy in the form of gamma rays. To match this feat, 100 trillion kilograms (110 billion tons) of dynamite would have to explode per second. However, nuclear fusion is such a thrifty energy source that only 0.034% of the mass involved in the reaction is truly lost.

Hydrogen fusion powers all stars in the initial stages of their lifecycles. Cool, eh? (Sorry, we meant hot!) Now, how do those gamma rays make the journey from the core to the Sun's 'surface'?

Given the crowded nature of the solar interior, gamma-ray photons are continually bumping into subatomic particles as they transit to the surface. If you have tried walking from one place to another in a very crowded room, you know the feeling. They are bumped this way and that as they slowly progress outward. Such 'random walks,'[3] occur many places in nature. If you open a bottle of perfume, you smell it first. Then, your friends progressively farther away smell it as the fragrance molecules make their way through the crowded air molecules to more-and-more distant noses.

With every collision, the gamma-ray photons lose some energy, so they gradually become x-ray,[4] ultra-violet, and then visible-light photons. Traveling at the speed of light, an unimpeded photon could reach the surface of the Sun in just a couple of seconds. However, because it undergoes a random walk, a photon may take from hundreds of thousands to a million years to reach the surface! This method of moving energy from one place to another is called radiative transport since radiation (photons) carry the energy.

At about 70% of the distance outward, the photon's energy catches a ride to the surface in rising hot parcels of gas, so only months may be required to complete the trip. In this region, convection moves the energy outwards (Figure 15.8).

In a close-up view of the Sun, we see can see 1,000-kilometer (600-mile) wide convective cells as a pattern of granulation. Larger-scale cells, 30,000 kilometers (20,000 miles) in diameter form supergranules. Nature always chooses the most efficient method to move energy from hot to cooler places (Figure 15.9).

When the photon reaches the 'surface,' its energy is mostly in the visible range. The light is now free to leave the Sun. The rest of the journey now takes only 8 minutes, 20 seconds to reach the Earth for sunbathers to enjoy.

Here is another interesting concept to consider. If the Sun were to explode now—it won't! It has got another good 5 billion years—we would not know about the event for an additional 8 minutes, since light from the Sun takes that amount of time to reach us.

Returning to those exotic neutrinos that are produced by hydrogen fusion, recall that we called them 'ghost' particles. They hardly interact with matter or photons. In other words, they travel

FIGURE 15.8   Boiling water is an example of convection.

FIGURE 15.9   Solar granulation.

unimpeded through walls, the Earth, or an imagined block of lead between the Sun and Earth with no problems. This 'antisocial' behavior makes them exceedingly hard to detect and measure. But think of the rewards of being able to peer directly into the core of our Sun nearly in real time. Through extraordinary means, astronomers have managed to do so, verifying our theory for what is happening in the invisible core of the Sun, known as the Standard Solar Model.

To catch neutrinos, scientists use instruments filled with liquids buried deep within the Earth, under the sea, or in large blocks of ice. The instruments are isolated to shield the detectors from unwanted other forms of radiation. American physicist Raymond Davis, Jr. ⟨1914–2006⟩ built the first neutrino 'telescope' using a 600,000-kilogram (660-ton) vat of carbon tetrachloride (dry-cleaning fluid) at the bottom of an abandoned mine shaft in Lead, South Dakota. On *extremely* rare occasions, a neutrino interacts with a chlorine atom, producing an atom of argon. Sensitive detectors count the number of argon atoms in the tank to estimate the number of neutrinos that have passed through. Even though the capture is rare, the sheer number of available neutrinos is *exceedingly* large. About 65 billion pass through an area the size of your fingernail every second!

As the South Dakota experiment proceeded, astronomers measured fewer neutrinos than theoretical predictions. It was always 1/3 the expected amount. Japanese physicist Masatoshi Koshiba ⟨1926–2020⟩ achieved similar results at the Kamioka Observatory.[5] Either physicists did not understand the nature of the Sun's core, or the experiment was not functioning properly. Each camp dug in and accused the other of mistakes, a controversy known as the Solar Neutrino Problem.

Further experiments performed by a Japanese research team led by Takaaki Kajita at the Super-Kamiokande detector located in the Japanese Alps and by a Canadian research group led by Arthur McDonald at the Sudbury Neutrino Observatory showed that neutrinos change their identities.[6] Three flavors of neutrinos occur in nature. One type, the electron neutrino, is produced in the Sun. During its journey to the Earth, it can 'oscillate' into one of the other types. On average, 1/3 of the solar neutrinos change into each of the other types and 1/3 remain unchanged, which resolves the puzzle scientists had wrestled with for decades. Imagine a magician who can convert a strawberry milkshake into a chocolate or vanilla one; neutrinos can do this by themselves!

This neutrino conversion also implies that neutrinos have mass and, therefore, travel slightly slower than the speed of light. In doing so, the two teams built on earlier work related to understanding neutrinos.[7]

It is hard to imagine that anybody ever will visit the deep interior of the Sun. The Standard Model of how the Sun operates is based upon two technologies that grew out of World War II: large machines that force subatomic particles to collide and computer simulations, the parameters of which are tweaked until their results match observations. It is to those observations that we now attend (Figure 15.10).

STARING AT THE SUN FOR EVEN SHORT PERIODS WILL DAMAGE YOUR EYESIGHT. As we *glance* at the Sun, we see a yellowish-white disk. This is the solar photosphere, meaning 'sphere of light,' since this is where visible light is released from the Sun. It appears to be a sharp surface or boundary, but it is much like walking in a fog. We can see near distances, but further down the path, we see what looks like an impenetrable opaque layer of white. The photosphere is simply the deepest we can look into the Sun. This is why we have insisted upon putting the word 'surface,' used in regard to the Sun, within quotation marks. The photosphere is only a few hundred kilometers (miles) thick. It often is referred to as the first layer of the 'solar atmosphere' (Figure 15.11).

Beginning with the ancient Greeks, and pretty much up to the time the telescope appeared, the Sun was supposed to be pure and unblemished. What a revelation it was then, when Galileo observed

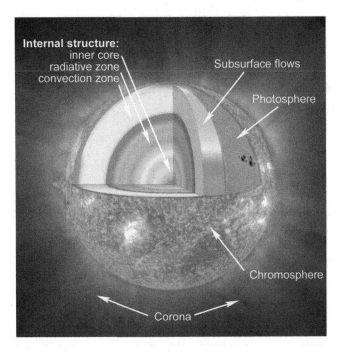

FIGURE 15.10 Diagram showing the layers of the Sun.

FIGURE 15.11   Railway line and row of telephone poles disappear into a fog.

dark markings on the photosphere, blotches on the pristine face of the Sun. OBSERVING THE SUN THROUGH ANY UNFILTERED TELESCOPE OR BINOCULARS IS UNSAFE (Figure 15.12).

These sunspots are not black. If we could lift one up and view it against the background of space, the sunspot would glow. However, at 1,700°C (3,000°F) cooler than the rest of the photosphere, sunspots do not emit as much light as their surroundings. To make the rest of the photosphere viewable given the ability of the human eye to register contrast, our eyes set sunspots to black, which they might as well be compared to the brilliant photosphere.

Sunspots start out as small 'pores' on the Sun. Most of these die out, but the survivors become full-fledged sunspots of such a size that one could easily dump the entire Earth into some of them. They last from a few days to a few months. They are characterized by a dark central solar

FIGURE 15.12   Sunspots.

photosphere surrounded by a less dark solar photosphere. Sunspots often occur in pairs or groups called active regions.

What are sunspots? They are regions of intense magnetic fields (thousands of times stronger that the solar average), twisted by the Sun's differential rotation. They suppress and divert rising heat from the interior. The result is local cooling.

In the 19th century, a German apothecary and amateur astronomer named Heinrich Schwabe ⟨1789–1875⟩ undertook counting the number of sunspots he could see. That was it. Just counting. Nevertheless, he discovered that during some years, sunspots were plentiful, but during other years they were scarce. Schwabe kept adding them up, and, later, others joined the effort. Eventually, a periodicity emerged from the sunspot count. A maximum number of sunspots are visible every (about) 11 years, expected to next occur in 2025. Similarly, sunspots are rare more-or-less halfway between sunspot count peaks (in 2020), thus following a corresponding period between sunspot troughs. A given cycle begins with spots at high latitudes; as it continues, spots appear closer to the solar equator (Figure 15.13).

The sunspot cycle continues today. The frequency of several phenomena associated with the Sun follows the sunspot cycle. These include the familiar planetary aurorae and solar prominences, which we will describe below. Because sunspots are highly magnetic, the sunspot cycle hints that other solar features are magnetic phenomena, too.

You might think that the Sun would dim with many sunspots on its surface. However, the light debt from sunspots often is balanced by the simultaneous appearance of brighter solar regions.

On top of the photosphere is a layer of the Sun that is invisible to us most of the time. Only during a total solar eclipse, when the light of the bright photosphere is blocked by the Moon, can we see the delicate chromosphere. It is only a few thousand kilometers (600 miles) thick. Its name, 'sphere of color' comes from its red tint, which is characteristic of hot, thin hydrogen gas. This glowing hydrogen indicates that the chromosphere is hotter than the photosphere. The structure of the chromosphere is like a forest of gaseous spikes, called spicules. Each spicule is several hundred kilometers (60 miles) wide and several thousand kilometers (600 miles) tall (Figure 15.14).

Professional astronomers now can block the light of the photosphere using an artificial occulting disk within their telescopes or a special filter that only transmits the light of the chromosphere. Using such coronagraphs, they can observe this delicate layer at will. However, in the past, they were willing to undertake long and lengthy expeditions to observe the chromosphere during a total eclipse.

Hovering over the photosphere and chromosphere is the amorphous corona. It is more rarefied than either. It is so hot, though, that some of its heavier constituent atoms have been stripped of many electrons. The corona may be several solar diameters wide but was invisible to us for the

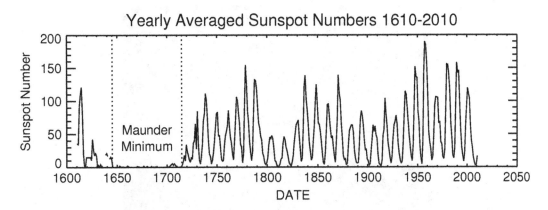

FIGURE 15.13 Graph of the sunspot cycle. Notice the mysterious absence of sunspots during an extend period in the 17th century.

FIGURE 15.14 Sun's chromosphere during a total solar eclipse.

same reason the chromosphere was: the bright photosphere. The corona remains a great draw to eclipse enthusiasts, as well. Its shape seems to change between sunspot maximum (round) and minimum (more butterfly-shaped) (Figure 15.15).

The temperature of the chromosphere is tens of thousands of degrees, and that of the corona is more than a million C (°F), which brings up a mystery: Hot travels to cold, right? Why then has the temperature of the Sun steadily decreased from the center, only to heat up again in the chromosphere and corona? The question still is debated. Is it 'heat bombs' emerging from within the Sun? Is it 'tornadoes' or waves in the chromosphere?

One possibility has to do with the fact that the solar 'engine' has a knock in it! The Sun vibrates. Inner layers slamming into outer layers may increase the temperature of the outermost solar atmosphere. Maybe.

FIGURE 15.15 Solar corona during the 2019 total eclipse of the Sun.

Even though the corona is extraordinarily hot, it is also extremely tenuous. An astronaut floating in the corona might encounter a million-degree atom on his forehead, and then another on his knee. However, the combined heat transferred to him would be so small that he still would freeze to death without his suit heater.

The corona is home to the most beautiful of solar features, the prominences. Prominences seem to erupt from the chromosphere like red-flame tongues that last for days or months. Often, they appear as bright loops. In the photosphere below a prominence, sunspots with opposing poles in of the local magnetic field may be seen. The prominences seem to follow magnetic field lines, hence their arch-like appearance—gaseous vaults big enough to envelop the entire Earth (Figure 15.16).

Leaving the Sun at hundreds of kilometers per second (miles per second) is the solar wind, permeating the inner Solar System and responsible for shaping the graceful tails of comets. As the solar wind flows unimpeded through the Solar System, it creates a vast bubble-like region surrounding the Sun known as the heliosphere. The boundary of this 'bubble,' called the heliopause, is formed when the solar wind with its embedded magnetic field encounters interstellar material permeating the Milky Way and the two mix. The heliopause marks the boundary between matter originating from the Sun and matter originating from the rest of the Galaxy. Outside the heliosphere, all is blowing in the interstellar wind!

The overall shape of the heliosphere resembles that of a comet. It is approximately spherical on one side out to about 120 au, with a long tail in the opposing direction, known as the heliotail, which stretches for several hundred au. The heliosphere partially shields the planets inside from dangerous cosmic rays and other radiation. NASA's Interstellar Boundary EXplorer mission imaged energetic neutral atoms from Earth orbit in 2008; these are mainly produced in a very narrow ribbon at its edges.

Five spacecraft have explored the outer heliosphere, including Pioneer 10 and 11, the twin Voyagers, and New Horizons. The two Voyager spacecraft traversed the heliopause in 2012 and can be said to have reached interstellar space. With another 50,000 au or so to go before they enter the Oort Cloud, they will leave the Solar System in several 10,000s of years!

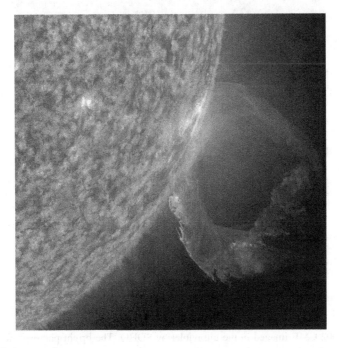

FIGURE 15.16   Solar prominence.

What is all this to us? Occasionally eruptions called flares and Coronal Mass Ejections [CME] hurtle trillions of kilograms (billions of tons) of materials through interplanetary space at incredible speeds.

If a CME encounters the Earth, it can trigger a geomagnetic storm that overpowers our magnetic field and the upper atmosphere. In addition to enhanced auroral activity, it also can induce electrical currents in the ground to disastrous effect. It can disrupt global radio communications, damage the fleet of Earth-orbiting satellites, and treat those flying in high-altitude passenger airplanes to a 'free' chest X-ray many times over. Outbursts of solar radiation imperil astronauts unlucky enough to be in space at the time, even though they may shelter in a special compartment with additional shielding.

For example, in 1989, a geomagnetic storm disrupted the power grid in Quebec, causing a 9-hour blackout. The associated aurorae were visible as far away as the southern USA. The economic impact of such an event can be in the billions of dollars.

Fortunately, we may have some warning time to prepare, ranging from hours to days. Space weather attempts to predict most urgently when a large eruption may occur on the Sun. By 'large' we mean a CME that sends forth a *trillion* kilograms (billion tons) of hot plasma in our direction! A whole new discipline has evolved: space meteorology.

We all have in our minds an image of the smiling Sun, making warm sunny days and allowing farms and gardens to grow (like on the Raisin Bran package), but clearly there is another, sinister side to the Sun. What if our eyes had not evolved beneath the light of the bland yellow photosphere? What if we had X-ray vision made famous by Superman? We would see the Sun as a horrible maelstrom. It definitely would not appear on a breakfast-cereal box.

Why study the Sun? Because it can kill you! (Figure 15.17)

To this end, solar physicists have sent numerous space probes to watch the Sun and its nearby environment. NASA, ESA, and the Canadian Space Agency deployed the Ulysses solar orbiter via the Space Shuttle in 1990. It headed first to Jupiter to get an orbital boost that sent it high above the ecliptic and back to the inner Planetary System, where it imaged the Sun's polar region for the first time and studied the solar wind at high latitudes.

2000/02/27 07:42

FIGURE 15.17   2002 CME imaged in the ultraviolet by SOHO. The bright photosphere has been blocked out. SOHO orbits the first Sun-Earth Lagrange Point.

In 1995, ESA and NASA launched the highly productive SOHO mission with instrumentation to study the Sun from core to corona to solar wind. Originally planned as a two-year mission, it has operated for more than 25 years, providing real-time observations for space weather predictions. NASA dispatched the Advanced Composition Explorer to detect high-energy particles in interplanetary space, most coming from the Sun. Installed in heliocentric orbit in 1997, it has successfully gathered critical data about the composition of the solar wind and its stormy nature.

JAXA launched Hinode[8] in 2006 to investigate the cycle of the Sun's magnetic field. It has provided essential clues for the causes of solar eruptions, their relation to the intense heating of the corona, and the driving mechanisms of the solar wind. The same year, NASA posted the twin Solar TErrestrial Relations Observatories [STEREO], one probe positioned at the Lagrange point ahead of the Earth in its orbit and the other behind. From these two vantage points, these twins produce stereo views of the Sun and view the region directly behind the Sun as seen from the Earth.

NASA stationed the Solar Dynamics Observatory [SDO] in 2010 to study how space weather affects our lives. Specifically, SDO is revealing how the Sun's magnetic fields are generated and how this magnetic energy is released to space. All these missions give important input to the Space Prediction Weather Center to warn of approaching geomagnetic storms.

Now two recent missions bravely approach the Sun. The Parker Solar Probe[9] [PSP] (NASA) was sent out in 2018 and is performing observations of the Sun's corona. On its final three orbits PSP will set a record for proximity to the Sun. It will fly within six solar radii above the Sun's 'surface,' more than seven times closer than the German/NASA Helios 2 spacecraft in 1974. The Parker Solar Probe successfully passed Venus on its way to the Sun in 2020.

ESA and NASA sent forth the Solar Orbiter in 2020, which will image the Sun's poles and study how the Sun creates and controls the heliosphere. The combined goals of these two space probes are to determine where the solar wind with its magnetic field originates, how its violent eruptions are produced, and how the Sun generates its magnetic field (Figure 15.18).

FIGURE 15.18   This is what the Sun would look like if we could see it in its X-ray emission. A flare bursts forth on the left limb.

No fundamental differences exist between the Sun and all the stars we see in the sky. The practical distinction is that because the Sun is so close to us, we can study it in detail. The Sun is our star. Are other stars the suns of other technological beings? Do other-world astronomers study our Sun from afar as we do theirs? As astronomers in the last century began to discover prodigious numbers of planets orbiting other stars, the possibility of answering this question increases.

NOTES

1 Or, the number "four" followed by 24 zeros, a truly astronomical number.
2 Gamma-rays are a form of electromagnetic radiation with wavelengths less than 0.01 nanometers (3.9×10^{-10} inches), much shorter than our eyes can detect.
3 Also known as a 'drunkard's walk,' for its resemblance to someone stumbling home from a bar.
4 X-rays are a form of electromagnetic radiation with wavelengths between wavelength range between 10 and 0.01 nanometers (3.9×10^{-7} and 3.9×10^{-10} inches), which are longer than gamma-rays but still much shorter than our eyes can detect.
5 Davis and Koshiba shared the 2002 Nobel Prize in Physics for the detection of cosmic neutrinos.
6 For this work, Kajita and McDonald shared the 2015 Nobel Prize in Physics.
7 Previously, American physicist Frederick Reines ⟨1918–1998⟩ pioneered experiments that confirmed the existence of neutrinos, for which he was awarded the 1995 Nobel Prize in Physics.
8 Meaning sunrise in Japanese.
9 Named in honor of American physicist Eugene Parker, who performed fundamental work on the Sun's wind. It is the first space mission named after a living person.

16 Alien Worlds

Are we alone? For life to exist as we know it, a planet with the proper conditions is a requirement. So as a first step to answering this question, we need to know if other planets exist in our Universe, or nearer, our home galaxy, the Milky Way.

No less a thinker than Aristotle thought that the Earth was unique and that we were alone in the Universe. But then again, he thought that the Earth was the center of the Universe. Another Greek philosopher, Epicurus of Samos ⟨341–270 BCE⟩, begged to differ and made the prescient statement, "There are infinite worlds both like and unlike this world of ours." Italian friar and mystic Giordano Bruno ⟨1548–1600⟩ amplified this sentiment in his own work, in which he speculated on infinite worlds like ours moving through infinite space (Figure 16.1).

Scientists, philosophers, and theologians continued to debate the existence of worlds other than our own and their possible inhabitants. Many astronomers expected planets should be common throughout space, if only they could be detected. In the late 20th century, astronomers settled the first question but left the second open.

What is an exoplanet?[1] Aside from the definition of a planet in our Solar System ➤ Chapter 2 ➤, an IAU commission defines an exoplanet as an object that orbits a star and has a mass less than that of 13 Jupiters. Curiously, no further criteria are given such as a lower mass limit, e.g., extrasolar asteroids and comets. The cores of these objects do not get hot or dense enough to support nuclear reactions.

More massive bodies up to 80 Jupiter masses are called brown dwarfs, objects intermediate between planets and stars. Brown dwarfs have sufficiently hot, dense cores that can sustain limited fusion reactions involving the rare isotope deuterium. African-American astronomer Gibor Basri, an authority on brown dwarfs, and Michael Brown, the discoverer of Eris, have explored the exoplanet-brown dwarf boundary for extrasolar systems (Figure 16.2).

Bodies having greater than 80 Jupiter masses are full-fledged stars since they can achieve core temperatures and pressures sufficient to initiate hydrogen fusion and generate their own light. Sustained nuclear fusion is the primary distinction between stars and planets.

Prior to the 1990s, the only known planetary system was our own. Detecting exoplanets is a daunting task. You might think astronomers should simply point their telescope at a host star, crank up the magnification, and snap a photo. Only it is not that simple. Remember, planets do not generate their own light and only shine by reflected starlight. Imagine trying to view a firefly flitting around a raging bonfire that is a billion times brighter and is a few thousands of kilometers distant. You get the idea: direct imaging is very challenging. Therefore, astronomers and instrument designers have had to develop indirect methods of finding exoplanets.

Astronomers use five main techniques to locate exoplanets: direct imaging, astrometric, radial velocity, transit, and microlensing. The four indirect methods identified the first planets orbiting other stars. However, doing so required advances in cutting-edge telescope and detector technologies that were unavailable prior to the 1990s.

Recall that two orbiting bodies move about a common center-of-mass with the more massive body nearest to this common point. ➤ Chapter 3 ➤ As a star moves through space, such orbital motion should appear as a slight wobble in its (transverse) path across the sky. Since the mass of the host star far exceeds that of the planet, the shifts are minuscule. This is known as the Astrometric[2] Method because astrometry is the branch of astronomy that measures positions and motions of stars.

Starting in the 1960s, Dutch-American astronomer Piet van de Kamp ⟨1901–1995⟩ reported a tantalizing wobble in the motion of Bernard's Star[3] as it moved across the sky, presumably caused

FIGURE 16.1   Epicurus shown in Roman reproduction of a Greek bust.

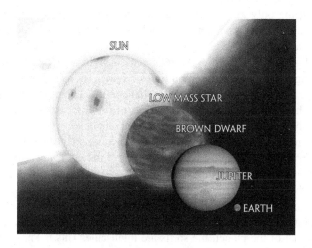

FIGURE 16.2   Size comparison for a brown dwarf.

by unseen low-mass bodies, i.e., planets. However, the wobble was always at the edge of the telescopes' discernable limits; it did not meet the strict confidence level needed to be generally accepted. Other groups observed Barnard's Star using multiple methods with mixed results. In 2018, a research team[4] reported a possible planet slightly more than three times the mass of the Earth orbiting Barnard's Star. Professed astrometric wobbles of other stars were put forward but not substantiated as exoplanets either.

Looking for a stellar wobble *is* a valid technique with which to detect exoplanets. Astronomers still use this today because advanced technology makes it feasible. ESA's Gaia, a robotic spacecraft orbiting the Sun, originally used this method. However, at the end of the 20th century, the Astrometric Method was sensitive to low mass stars, rather than planets, orbiting other stars (Figure 16.3).

Moving through the galaxy, a star's velocity also has a radial component, either towards us or away from us. Therefore, astronomers can also alternatively measure the minute wobble in this motion caused by the orbiting exoplanet. Sensitive instruments, called spectrometers, spread starlight into its component colors revealing slight shifts towards shorter and longer wavelengths. Astronomers can measure the star's varying speed and direction from the magnitude of this shift as the star alternately approaches then recedes from us. This Radial Velocity Method is another application of the Doppler Effect. ➤ Chapter 6 ➤ The technique is biased towards massive planets circling very close to their parent stars, inducing the largest shifts, and to nearby bright stars. The approach is responsible for about 1/5 of exoplanet discoveries. As a bonus, astronomers can usually infer the planet's mass and distance from the host star using data obtained by the Radial Velocity Method (Figure 16.4).

Mayor and Queloz used the Radial Velocity Method when they finally resolved the question of whether exoplanets exist. Recall that, in 1995, they discovered 51 Peg b orbiting an ordinary star. ➤ Chapter 1 ➤ It is a Jupiter-mass planet rounding its ordinary star closer than Mercury orbits the Sun, a type now known as a 'hot Jupiter' (Figure 16.5).

51 Peg b? The nomenclature for exoplanets is straightforward. Start with the star's name from a catalogue (51 Peg) or discovery telescope. Then, append a letter starting with a lower-case *b*

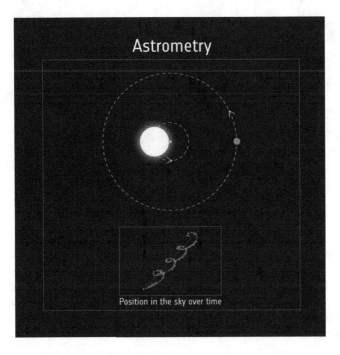

FIGURE 16.3   Diagram of the Astrometric Method.

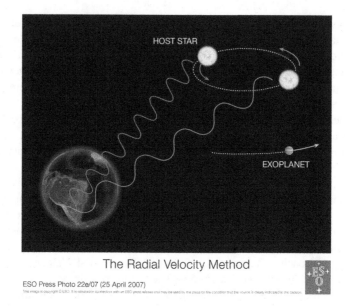

The Radial Velocity Method

ESO Press Photo 22e/07 (25 April 2007)
This image is copyright © ESO. It is released in connection with an ESO press release and may be used by the press on the condition that the source is clearly indicated in the caption.

FIGURE 16.4 Diagram of the Radial Velocity Method.

FIGURE 16.5 Artist's rendering of 51 Pegasus b (Bellerophon).

(51 Peg *b*); the letter *a* is reserved for the host star. For additional planets, increment the letter in order of discovery, regardless of distance from the star. Hence, Earth would be designated *Sol g* in this scheme, since it was the sixth planet to be recognized as such in our Solar System.

In fact, two planets had already been found in 1992. However, these orbited a neutron star, which is the remains of a high-mass star after a violent supernova explosion. Although intriguing, this discovery disappointed many astronomers, whose quest was finding planets around stars like our own.

Earlier we pointed out the difficulties associated with direct imaging. Ordinarily, the planet is lost in the glare of the host star. To overcome this problem, a special instrument called a coronagraph is attached to a telescope. Mounted in it is an opaque disk that blocks the light from the star revealing its faint surroundings. This technique also is used to find proto-planetary disks and to study the Sun's corona. In addition, a special system called adaptive optics can sharpen the image that would otherwise be blurred by the Earth's atmospheric motions. Observing in infrared light also improves the situation, since at those wavelengths the contrast between star and planet is less than in visible light. Although discovery is still very difficult, direct imaging of exoplanets became possible in 2004, at least for some of the nearest stars. However, this method has made only about 1% of exoplanet discoveries (Figure 16.6).

A very productive technique for finding exoplanets is the Transit Method. Like transits of Mercury and Venus in our Solar System, this approach requires a planet to pass alternatively in front of and then behind the star it orbits. The plane of the orbit must be nearly in our line of sight for this to occur. When the planet transits its star, the light of the star is slightly diminished. Accurately measuring this tiny dimming over time produces a light curve. Analyzing the light curve, astronomers can determine the planet's orbital period and diameter. Combining size with mass (determined by the Radial Velocity Method), the planet's density can be calculated, which provides clues to its composition (i.e., whether it is a rocky terrestrial body or a mostly gaseous one like Jupiter). With over 3,000 transiting planets detected (3/4 of the total to date), it is by far the most fruitful method (Figure 16.7).

Both ground- and space-based telescopes, including small amateur telescopes equipped with sensitive light-measuring detectors,[5] used the Transit Method. The French Convection, Rotation, and planetary Transits Mission [CoRoT] was one such small space telescope that discovered 32 exoplanets during its six-year operation. Launched in 2009, NASA's Kepler Telescope employed the Transit Method and has been the workhorse of exoplanet research. Designed to find low-mass, Earth-like planets, it discovered more than 3,000 confirmed exoplanets, with an additional 2,000 or so awaiting confirmation.

The Transiting Exoplanet Survey Satellite [TESS], launched in 2018, is NASA's next step in the search for planets outside our Solar System. Replacing Kepler, TESS will search for transits in

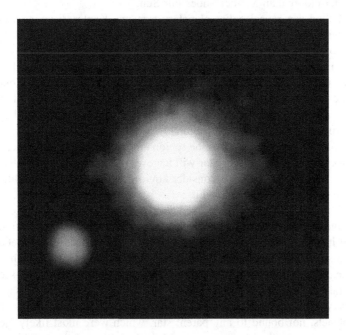

FIGURE 16.6 Direct imaging of an exoplanet only became possible in 2004.

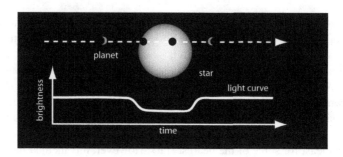

FIGURE 16.7   Diagram of the Transit Method.

200,000 of the brightest stars near the sun. It has discovered 55 exoplanets to date, with over 2,000 others awaiting confirmation.

Other missions have joined the race to find transiting exoplanets, including ESA's CHaracterising ExOPlanets Satellite [CHEOPS][6] launched in 2019 and the newly repurposed Gaia, which launched in 2013. The specialty of CHEOPS is measuring exoplanet diameters precisely to reveal their densities.

In addition, the Hubble and Spitzer Space Telescopes have been employed in the search for and study of exoplanets. Two highly anticipated NASA facilities, the James Webb Space Telescope[7] and the Nancy Grace Roman Space Telescope[8] will devote part of their observing time to hunt for exoplanets also.

Specialized ground-based telescopes have joined the exoplanet rummage: Hungarian-made Automated Telescope Network [HATNet] Exoplanet Survey (2001), High Accuracy Radial velocity Planet Searcher [HARPS] (2003), Wide Angle Search for Planets [SuperWASP] (2004), Automated Planet Finder [APF] (2013), Next-Generation Transit Survey [NGTS] (2015), the automated TRAnsiting Planets and PlanetesImals Small Telescope [TRAPPIST] (2010), and others. The Kilodegree Extremely Little Telescope [KELT] made its first transit discovery in 2017. KELT-11 b is a puffy gas giant; its radius is 1.5 times that of Jupiter, its mass 1/4 that of Jupiter. Yet, it orbits its star closer than Mercury does our Sun.

The Transit Method can reveal entire planetary systems. Kepler discovered eight Earth-mass planets orbiting within the 'Habitable Zone' of their star. Recently, TRAPPIST netted seven terrestrial planets orbiting a cool red dwarf star now known as TRAPPIST-1. Five of these exoplanets are Earth-sized but the other two fall between Earth and Mars in size. All are likely locked tidally to their parent star. The three outermost lie within the star's Habitable Zone, including one whose atmosphere contains water vapor. To an astronaut standing on one of these, the others would appear prominently in her night sky. She should be cautious to avoid the powerful X-ray flares known to erupt from the host star! (Figure 16.8)

What is the Habitable Zone? It refers to the region around the parent star where water can exist as a liquid. A planet orbiting too near its star will have all its water vaporized while, if it is too far, its water will be frozen as ice. Biologists consider liquid water an essential condition for life as we know it, hence the use of the word 'habitable'.

Microlensing is the final technique employed thus far in the search to find exoplanets. Einstein proposed that gravity deflects light much like mass. So, when an exoplanet transits, its gravity bends the light of its parent star, causing a brief brightening. In effect, the planet acts like a small lens, focusing the star's light, thus the moniker microlensing. A measure of the flash yields the mass of the planet. An advantage of this approach is that the flashes can be seen at very large distances. Using this technique, astronomers have found evidence of exoplanets as far away as the Andromeda Galaxy, some 2.5 million light years distant! Microlensing also has found free-floating, rogue planets, not bound to any parent star, which were most likely ejected from their birth systems. So far, this method has discovered less than 100 exoplanets (Figure 16.9).

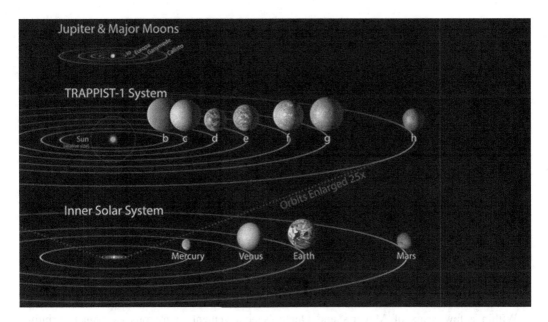

FIGURE 16.8 Diagram comparing the TRAPPIST 1 system to the inner Solar System (to scale).

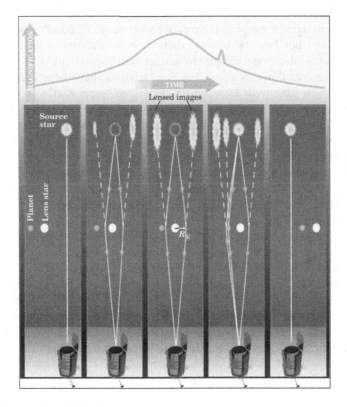

FIGURE 16.9 Microlensing Method.

FIGURE 16.10   Kepler 62 f, rendered here, is a super-Earth.

Within a few years of Mayor's and Queloz's achievement, numerous exoplanets orbiting normal stars were found. To date, astronomers know more than 4,000 such objects with a wide variety of orbits and characteristics. Many exoplanets have been found to possess atmospheres, rings, moons, and magnetic fields; some are in many ways Earth-like.

A visit to the menagerie of exoplanets reveals a wide range of possible worlds. Initially, astronomers found many 'hot Jupiters' because their characteristics make them easy pickings for several detection techniques. Some exoplanets are termed 'super-Jupiters,' since they approach 13 Jupiter masses. A notch down in mass, 'hot Neptunes' also have been found.

As detection methods improved, astronomers discovered 'super-Earths' with masses between that of Earth and Jupiter, which are now the most numerous kind known. A few very massive rocky planets more than ten times that of Earth have been located, termed 'mega-Earths.' In addition, gaseous bodies with masses between about 2–10 Earth masses called 'mini-Neptunes' were observed (Figure 16.10).

Eventually, the precision was achieved to detect Earth-like, rocky terrestrial planets. Some Earth-mass worlds reside in the warm Habitable Zone where liquid water can exist on their surfaces. Some exoplanet atmospheres have familiar components: water vapor, carbon dioxide, and methane. Might living things be present, too?

The study of exoplanets is mature enough to report exoplanets with moons ('exomoons') and evidence of 'exocomets.' Of the more than 4,000 exoplanets found to date, about 3,000 are found in systems with more than one planetary member; roughly 700 are in multi-planet systems with more than two planets. The current record is eight, tying our Solar System (Figure 16.11).

Exoplanet searches have revealed a remarkable variety of exoplanets and planetary systems to which they belong. Keeping with the theme of this book, most are worlds of ice and fire. To illustrate the range of strange and bizarre exoplanets, we describe five fascinating realms. Remember, these are real worlds, not something from science-fiction books or movies.

Kepler-7 b is a large world, 50% larger than Jupiter but only 1/2 its mass, resulting in a density of only 0.14 g/cm³. Like Saturn, it would float on water! It has been nicknamed the Styrofoam® world (Figure 16.12).

An Earth-sized world, Proxima Centauri b, revolves about the next nearest star, Proxima Centauri, a cool, red star. The planet has a mass 1.3 that of the Earth, orbits its host in 11 days, and is in the Habitable Zone. Breakthrough Starshot is a private project to travel to this system proposed by Russian physicist Yuri Milner, English physicist Stephen Hawking 〈1942–2018〉, and

FIGURE 16.11   Artist envisions Kepler 90's eight planets.

FIGURE 16.12   Artist compares Kepler-7 b to Jupiter.

American entrepreneur Mark Zuckerberg. The ultra-low-mass (grams), very small (centimeters) probe would utilize the pressure of sunlight for propulsion, or 'light sail' technology. A high-power laser cannon mounted on the Earth would target the probe, providing the light pressure. Achieving a speed around 20% the speed of light, the miniature robot would arrive in *only* 20 to 30 years. After which, four more years would lapse to transmit the return signal, providing the next generation of astronomers with images and other information (Figure 16.13).

Ogle-2005-BLG-390L b is a cold planet five times more massive than the Earth. This super-Earth has a surface temperature of −220°C (−360°F). Nicknamed 'Hoth,' after the fictional *Star Wars* planet, it may be a failed mini-Neptune. Talk about an ice world! (Figure 16.14)

Another super-Earth is Kepler-22 b, orbiting a Sun-like star in 290 days. It has a diameter 2.4 times that of the Earth and a Uranus-like obliquity. The average temperature is a comfortable 22°C (72°F), and it may have a super ocean covering its surface that keeps its climate mild. Nicknamed 'Kamino,' after another fictional *Stars Wars* world, it seems to be an example of a 'Goldilocks' planet (Figure 16.15).

FIGURE 16.13 Artist's impression of Proxima Centauri b.

FIGURE 16.14 Artist's depiction of Ogle-2005-BLG-390L b (Hoth).

Ever wonder what it is like in Dante's lowest ring of Hell? Look no further than Wasp-76 b, an ultra-hot Jupiter. Its uniquely-polar orbit gives it a period of 2 days, and it is tidally locked to its parent star. Dayside temperatures soar to 2,400°C (4,350°F), high enough for metals like iron to be vaporized. Nightside is cooler—'only' about 1,500°C (2,700°F). Strong winds carry the iron vapor from the dayside to nightside. The iron condenses into clouds when it reaches this darker, cooler hemisphere of the planet and, then, precipitates as rain consisting of extremely hot iron vapor droplets. It literally rains liquid iron on the nightside. Try to match this one for a fire world! (Sorry Io!) (Figure 16.16)

New space missions and ground-based facilities (e.g., 30-m class telescopes and pairs of telescopes with extremely high resolution) will open a new era in exoplanet studies and will revolutionize further our picture of them in the coming years. Detections of atmospheres (e.g., chemical composition, including molecules indicative of living organisms and the presence of clouds), accurate mass and radii determinations, constraints on exoplanet composition and interior structure, rotation and orbits, day-to-night side differences, and the habitability of a large sample of exoplanets are all within the grasp of these powerful tools.

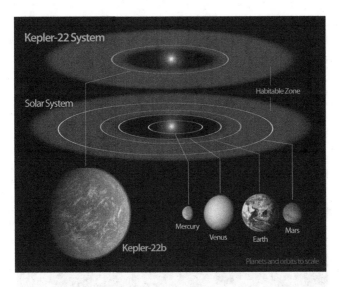

FIGURE 16.15   Diagram showing Kepler 22 b's (Kamino's) position in its planetary system compared to the Earth's location in the Solar System.

FIGURE 16.16   Artist's interpretation of WASP 76 b.

To understand the increasing number and variety of exoplanets, we also need to know about their parent stars. The star we know best is our Sun. ➤ Chapter 15 ➤ How does it, as the host of eight planets and billions of smaller bodies, compare with stars that are hosts of exoplanets?

When we gaze into the night sky, we see a multitude of distant suns with different brightnesses and colors. Their brightness is related to the star's temperature, size, and distance. Color is the astronomer's thermometer, revealing a star's temperature: the bluer, the hotter, and the redder, the cooler. British-American astronomer Cecilia Payne-Gaposchkin ⟨1900–1979⟩ developed the system used today to accurately determine stellar temperatures.

If astronomers can measure the distance to a star, they can find its size in combination with the temperature. Doing so at the beginning of the 20th century, Danish astronomer Ejnar Hertzsprung ⟨1873–1967⟩ and American astronomer Henry Norris Russell ⟨1877–1957⟩ found that most stars follow a relationship whereby the more intrinsically bright, the hotter and more massive, called the Main Sequence. In addition, other stellar groups also exist. Astronomers recognize that groups with common properties found by Hertzsprung and Russell represent stars in various stages of their lifecycles (Figure 16.17).

The main-sequence stars are in the longest, most stable phase of the stellar lifecycle and represent stars with masses from 100 solar masses at the hottest end to a fraction of a solar mass at

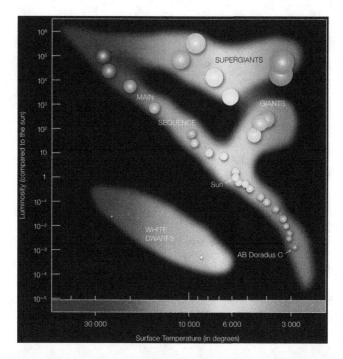

FIGURE 16.17   When Hertzsprung and Russell drew a plot of stellar luminosity versus temperature, they found that stars form groups such as the Main Sequence and Giants.

the coolest. The most massive stars on the Main Sequence have lifetimes of a few million years, whereas the least massive live 100 billion years or more. Stellar temperatures range from 50,000°C (90,000°F) associated with bright, blue stars to around 3,000°C (5,000°F) for the dim, red stars. Habitable Zones scale with stellar temperature and size; the hotter the star, the farther and more expansive is its Habitable Zone.

Putting the Sun in context with other main sequence stars, ours has average properties, including temperature, size, mass, brightness, and age. It lies towards the lower end of the main sequence with an effective surface temperature of 5,500°C (10,000°F) and an anticipated 10-billion-year lifetime.

Off the main sequence, more evolved stars exist, such as large subgiants (up to 10 times larger than Sun), giants (10–100 times larger than Sun), and supergiants (1,000 times the Sun's size), as well as merely Earth-sized, hot stellar remnants known as white dwarfs.

Based on chemical composition, stars generally can be grouped into two categories: older ones with little or no metals and younger ones with higher amounts. Our Sun is among those with higher metallicity. Having more raw materials certainly would be conducive to planetary formation.

Astronomers find that more than half of all stars have a stellar or brown dwarf companion. Our Sun is in the minority in this respect. Can planets have stable orbits in multiple stellar systems? Surprisingly, the answer is yes. Planets have been found in binary stellar systems, either closely orbiting one star or at a distance revolving about both. Planets have been detected in triple star systems, and even one planet is known in a quadruple system (Figure 16.18).

Detecting so many exoplanets in orbit around a diverse range of stars implies that planets are a common byproduct of star formation. When astronomers turn their telescopes to large gas and dust clouds in space, such as the Great Orion Nebula, they find proto-stars embedded in the clouds. These young stars are in their earliest stages of formation. The nebulae are virtual stellar maternity wards, the birthplace of stars. Surrounding the proto-stars are proto-planetary disks or proplyds, which are large flattened disks of gas and dust. Roughly 5 billion years ago, our Solar System

FIGURE 16.18   Artist's view of exoplanet Kepler 47 c orbiting a binary star.

probably looked the same. ⬾ Chapter 1 ⬾ Given time, we expect planets to form in the proplyds following the same general processes that formed our Solar System.

Astronomers also observe debris disks around young stars. Some of these disks of dust and other fragments are divided into distinct rings. This phase of development may follow the proto-planetary disk phase once terrestrial planets finish growing.

Eventually, the remaining gas and dust dissipates; a young planetary system is born. This would be a likely place to observe the development of young exoplanets. The problem is the time required for planetary formation, estimated to be at least millions of years.

Astronomers do not have time to watch a single proto-planetary disk waiting to see planets form. But by observing many proplyds, they expect to catch systems in a variety of stages of planet formation. The images can be assembled like puzzle pieces in a probable chronological sequence.

Studies of exoplanets and exoplanetary systems are important to understand our Solar System and to identify the planetary properties that would support the emergence of life as we know it. Astronomers have begun to link the architecture of planetary systems to the properties of the planet-forming disks around young stars. These efforts should eventually trace the present-day properties of these exoplanets (mass, diameter, chemical composition, atmosphere, etc.) to their formation history and evolution to maturity. Here, theoretical models that include the relevant physics and chemistry are important guides.

What can be learned about our Solar System by comparison with planetary systems found around other stars? How does the structure and history of exoplanetary systems compare with ours? Planetary systems most commonly orbit low- and intermediate mass main-sequence stars, like our Sun. Stars with higher metal content are more likely to have planetary systems, too. When astronomers examine the hundreds of multi-planet systems, they find that our Solar System is very different, though.

First, the spacing of the planets varies greatly. 'Hot Jupiters,' giant planets revolving very near their central star, are common, whereas our giants are located at a safe distance from our star. Our small terrestrial planets are close in (Figure 16.19).

Second, as mentioned earlier, the most-common exoplanets are super-Earths. We have none in our Solar System. Surely as our technical abilities improve, astronomers will find many more, smaller terrestrial planets, but how will they compare with our terrestrials? Likewise, mini-Neptunes, mega-Earths, super-Jupiters, hot Neptunes, and others are strangers to our neck of the woods.

Third, although some exoplanet orbits are near circular, others can be highly elliptical, greatly inclined, and in retrograde directions, more like comets in our Solar System. Some transiting

FIGURE 16.19   Artists' depictions of 'hot Jupiters.'

planets orbit in the direction opposite to their central star's rotation. As we know, all eight major planets in our Solar System have near-circular orbits in roughly the same equatorial plane and travel in the same direction as the Sun's rotation. Exoplanet studies indicate that systems with a multiplicity of planets tend to have rounder orbits, whereas those with few planets have highly elliptical orbits.

Since astronomers do not think it possible that hot Jupiters could have formed so unexpectedly close to their host star, perhaps the mechanism that caused the inward migration also may have resulted in the unanticipated backwards or tilted orbits. With all of this reordering of planets, what is the importance of collisions early on in shaping planetary systems?

So many new questions have been raised about the formation of stellar systems that we need to go back to the 'drawing board' to understand the new data. Most likely we must revisit our Solar

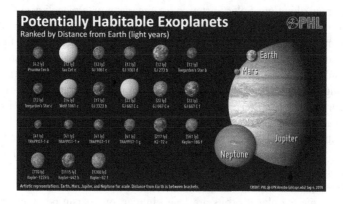

FIGURE 16.20   Artists' illustrations of potentially habitable exoplanets.

Nebula Theory, including the Nice and Grand Tack Models. ➤ Chapter 1 ➤ Although finely tuned to our Solar System, it is certainly lacking when we apply it to exoplanetary systems.

But this newly found knowledge is what makes science so exciting. It challenges us to apply our scientific methods and tools in new and novel ways to understand the worlds around us. Clearly this intriguing field of astronomy has a bright and interesting future!

In conclusion, our galaxy likely has a very large number of hospitable, Earth-size planets. Do intelligent beings inhabit any of these other worlds? Or are we the only ones available to ask such questions? We haven't a clue! (Figure 16.20)

NOTES

1 Also called an extrasolar planet.
2 From astro (star) + metry (measuring).
3 Named for American astronomer Edward Emmerson Barnard ⟨1857–1923⟩ who, in 1916, first reported its fast motion across the sky with respect to background stars.
4 Using radial velocity measurements (discussed later) in this case.
5 That is, modern charged-couple device [CCD] cameras.
6 An alternative English spelling for Khufu ⟨*circa* 2,600 BCE⟩, the 4th dynasty pharaoh for whom the Great Pyramid of Giza was built.
7 Replacement for HST scheduled to launch in 2021.
8 Replacement for Spitzer Space Telescope scheduled to launch in the mid-2020s.

17 Future Exploration and Adventure in Our Solar System

Let us continue by considering the question posed at the end of the last chapter: Are we alone? Or even the related question: Has life ever existed in the Universe in the past? The answer from the scientific viewpoint is simply: We don't know! But an optimist can find reasonable grounds to argue that the answer is: No, we are not alone. We just do not have the necessary tools to find life in this vast Universe, even in our neighboring planets. But we are close! We can now send advanced probes throughout the Solar System with instruments to search for life. However, we still need the political will to provide the necessary funding. Some entrepreneurs are not waiting and are tackling this question on their own.

The scientific search for life in the Universe and the necessary conditions for its origin is astrobiology.[1] It is synergetic with the study of the Solar System and exoplanets. ⇥ Chapter 16 ⇥ By necessity, astrobiologists must rely heavily on life as we know it. Perhaps some exotic unknown lifeforms exist based on, for example, phosphorus or hydrogen, instead of our familiar carbon chemistry. Instead of speculation, we prefer to concentrate on what is known.

We must also stress that our musings about life's origins are far from certain. The Earth retains many clues to the origins of life in its geologic and fossil records. However, biologists still do not understand completely how life came to be from a scientific viewpoint. With these limitations firmly in mind, let us proceed.

The basics of astrobiology state that three ingredients are required for life: organic (carbon-based) compounds, liquid water, and an energy (heat) source. The organics can be simple molecules, like hydrogen cyanide and formaldehyde. Prebiotic chemical reactions can transform these into the more complex ones required for life, such as amino acids, enzymes, ribonucleic acid [RNA], and DNA. Water acts as the perfect solvent for allowing these reactions to occur. Surfaces, such as dust grains, can act as catalysts to promote them. Energy is necessary to stimulate the reactions to proceed at a reasonable pace.

Which, surprisingly, brings us back to the study of comets. ⇥ Chapter 13 ⇥ Recall that according to the molecular panspermia theory, comets and asteroids transport water and prebiotic organic compounds, basic life ingredients, to the Earth via impacts. Biologists do not think that life arises in these small bodies. Instead, life needs a solid planetary surface, preferably with an ocean or even just a 'warm little pond,' and maybe an atmosphere, too.

In 1996, scientists from NASA's Johnson Space Center made a startling announcement heard around the world. They claimed to have found microfossils in a rock from Mars!

The rock is Allan Hills 84001 (ALH-84001),[2] a martian meteorite collected in Antarctica and stored at the Johnson Space Center for a dozen years. Studies of this rock determined that it was of martian origin and about 4 billion years old. Some 17 million years ago, a violent collision blasted it from the planet's surface, forcing it to travel through space for millions of years. Then, about 13,000 years ago, it ventured close enough for Earth's gravity to bring it to rest on a different planetary surface. We are very fortunate to have about a dozen surface samples from Mars in a variety of meteorite collections around the world. We did not even need to travel to the red planet; they came to us for free! This history of ALH-84001 is undisputed.

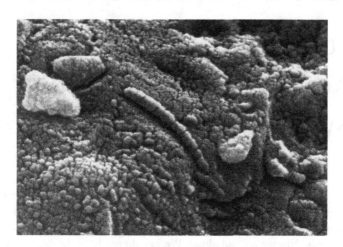

FIGURE 17.1   Interior of meteorite Allan Hills 84001. The tubular structures are smaller than 100 nanometers long.

Under a powerful microscope, scientists saw worm-like structures resembling microfossils
found on the Earth, only much smaller. Examination of ALH-84001 indicated that cracks had
formed in it while on Mars, and hot water had percolated throughout, heated by geological activity
billions of years ago. They interpreted this evidence to mean that colonies of nanobacteria thrived
on Mars until the rock was blasted into space (Figure 17.1).

The scientists ruled out contamination of ALH-84001 from Antarctica because the concentrations of organics were low at the surface and increased with depth into the rock. The distribution is
the opposite of what would be expected if contaminants entered at the surface and made their way
into the interior. Therefore, they concluded that this meteorite is evidence of life having existed on
Mars in the ancient past.

As American planetary astronomer and author Carl Sagan ⟨1934–1996⟩ once said, "extraordinary
claims require extraordinary evidence" to meet the high bar of acceptance. In the ALH-84001 case,
alternate non-biological explanations exist for each step of the hypothesis. So, the scientific community
has largely rejected this meteorite as evidence of martian life.

The affair highlighted the interesting fact that material can be transported from planet to planet
through collisions; life could come along for the ride. The Earth may have been invaded by
extraterrestrials, not alien beings in advanced craft but rather microorganisms inside meteoroids. In
fact, with recent evidence that some asteroids and comets are of interstellar origin, life could
potentially even travel here from other planetary systems! Of course, the opposite is also true:
Earth life may have traveled to Mars, seeding the red planet.

We understand how planetary material, perhaps containing fossils, could come to the Earth, but
could living ancient 'astronauts' inside the rocks survive the journey? It seems it is possible depending on the organism.

When biologists investigate extreme Earth environments, they see such locations teeming with
organisms known as extremophiles. Life is very resilient and adaptable to its environment. Some
forms thrive regardless of whether their home is extremely acidic (acidophiles) or alkaline
(alkaliphiles), hot (thermophiles) or cold (cryophiles), oxygen-free (anaerobe) or deep in the
Earth with no sunlight (endoliths). Life exists even in radioactive nuclear reactors that would kill
us in a matter of minutes (radioresistant). Some creatures share multiple characteristics (poly-
extremophiles). When exposed to the vacuum of space, many extremophiles become spores, highly
resistant, dormant organisms with no metabolic activity. They come 'back to life' when more
favorable environmental conditions exist. Such organisms are the most interesting candidates to be
micro-astronauts (Figure 17.2).

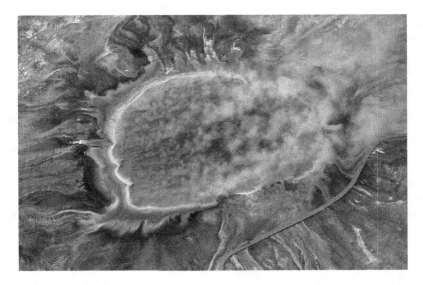

FIGURE 17.2 This pool in Yellowstone National Park is very hot. Notice the steam coming from it. Yet, thermophilic algae, bacteria, and archaea flourish as the colorful mat seen around the spring's edge.

Although these extremophiles are mostly single-celled bacteria, very sturdy and resilient multicellular organisms, like tardigrades,[3] also exist. You can find these eight-legged critters in a local pond and see them without microscopic aid. They can survive extreme temperatures and pressures, radiation, dehydration, and vacuums like those of space. Tardigrades have been frozen for years and when thawed, spring back to life! They are among the most resilient animals known. Tardigrades might be found on the Moon now that the private Israeli spacecraft Beresheet[4] crashed into the lunar surface with a load of them in 2019. Aside from this mishap, no evidence exists for life on the Moon (Figure 17.3).

Another strategy for finding extraterrestrial life is to search for atmospheric biosignatures, by-products of living organisms on the Earth. Such biosignatures include small amounts of methane, phosphine, or chlorophyll, or large amounts of oxygen. Astronomers are debating the significance of methane as a biosignature on Mars. ➤ Chapter 8 ➤ In addition, some astronomers have tentatively

FIGURE 17.3 Tardigrade (about 0.05 centimeter [0.02 inches] in length).

detected phosphine in Venus's atmosphere, but this 2020 result is disputed by others. If present, microbes living in the planet's clouds where temperatures and pressures are more conducive to life could produce the phosphine. However, before astrobiologists can attribute phosphine to microbes, the controversy of its presence must be settled. The scientific process is at work.

Searching for life is one thing, finding intelligent life is another. As captured in the fossil record, unicellular life appeared early, perhaps 4 billion years ago when the Earth was a spritely half-billion years old. Intelligent life came much later. Anthropologists place this event at about a few hundred thousand years ago.

Imagine the history of the Earth placed on a one-year calendar, with its origin occurring on January 1st. One second on the calendar represents 144 years. The first signs of life would occur in late February. The first multicellular organisms do not arise until mid-August. After that date, all manner of plants and animals appear with increasing complexity. Then, on December 31st, about mid-day, the first sign of intelligent life graces the Earth. Modern humans would enter the scene a few minutes before midnight! We are infants in the grand scheme of life on the Earth (Figure 17.4).

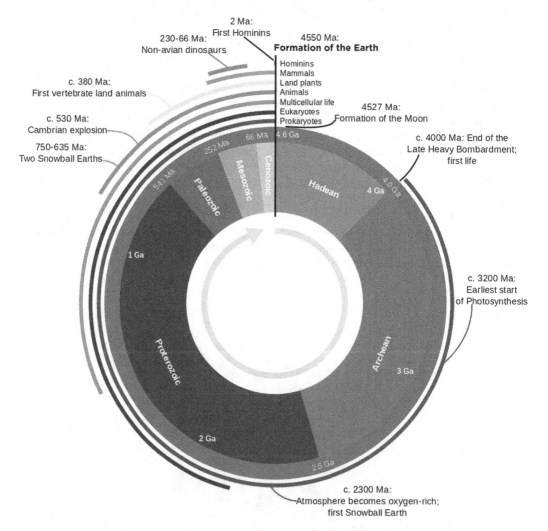

FIGURE 17.4 Here, Earth history is diagramed in a single rotation of a clock. Notice the relative times of the Late Heavy Bombardment, Snowball Earths, and wide distribution of different lifeforms called the Cambrian Explosion.

Although 'infants,' our species has initiated multiple serious searches for intelligent life elsewhere in the Universe. Such scientific investigations are collectively known as the Search for Extra-Terrestrial Intelligence [SETI]. Around the world, interdisciplinary projects combine astronomy, biology, geology, climatology, and other fields to investigate SETI-related topics. The SETI Institute, a non-profit research facility located in Mountain View, California, specializes in addressing questions about the development of life, where it may exist in the Universe, and how it might be discovered. As the saying goes, "The truth is out there!"; we only need to search for it.

Almost every week, the lurid press reports an unexplained event in the sky, an Unidentified Flying Object [UFO]. Are UFOs real? Under the strict technical definition, yes; investigators have not been able to identify unequivocally all apparently 'flying' objects reported.

Scientists have no ready explanation for some of these sightings, because analyzing anecdotal stories accompanied by insufficient details is difficult. However, jumping to the conclusion that these apocryphal stories prove that alien-powered craft are buzzing our Earth is irresponsible. Some unscrupulous individuals have milked spectacular claims of alien visitation or abduction for financial gain. We want to believe, so fraud is rampant! Equip yourself with critical thinking skills and a large dose of skepticism when evaluating such claims. Armed with basic astronomical knowledge about the position of Venus in the sky, you can 'solve' about 40% of reported UFO claims!

Astrobiologists are confident that no other intelligent life inhabits our Solar System, so SETI focuses on nearby stars, the Milky Way, and neighboring galaxies. American astronomer Frank Drake formulated a method to estimate the possibility that intelligent life exists in the Milky Way. Rather than being an exact calculation, the Drake Equation summarizes the main concepts to stimulate discussion about them (Figure 17.5).

Using it, we estimate the various factors that would favor intelligent life, such as the chance of finding a hospitable host star with a terrestrial planet in the Goldilocks Zone. We also estimate the fraction of habitable planets that develop intelligent life with the ability to communicate their existence and the lifetime of such civilizations. Obviously, many factors are highly conjectural or simply not known. Therefore, we must use educated guesses. As a result, the estimated number of advanced civilizations in the cosmos ranges from an optimistic number in the millions to a pessimistic one (just us)!

What guides the future of astronomical research, including planetary science and astrobiology? Is it every scientific team for themselves? Although astronomical research benefits from academic

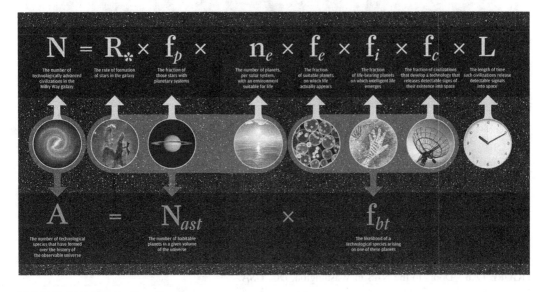

FIGURE 17.5 Drake Equation.

freedom, most scientists must obtain funding to pursue a hypothesis or experiment. Starting in the mid-20th century, large government-funded agencies like NASA and ESA provide most research funds, relegating independently wealthy benefactors to a minor role.

In the US, astronomers must compete for limited grants from these agencies by submitting proposals for their work—a new spacecraft mission, a new instrument, etc. Scientific peers review these proposals and recommend the 'best' ones for funding. The competition is fierce, with many funding programs oversubscribed by as much as a factor of ten! A similar peer-review system awards time at major telescopic facilities, as well.

To guide US policy decisions, the National Academy of Sciences [NAS][5] drafts the Decadal Survey on Planetary Science and Astrobiology[6]; it represents the consensus of the planetary science community about future priorities and challenges. Proposers should study this document before submitting their requests for funding! For the rest of us, it is an important roadmap of where astronomy is headed in the next and following decades (Figure 17.6).

NASA's "Follow the Water" strategy for Mars exploration grew out of the Decadal Survey. Looking for water ice and warm aquifers, searching for signs of ancient volcanic activity, and determining the evolution of the martian atmosphere all support the goal of discovering evidence of life on the red planet. In addition, the New Horizons mission to Pluto and the Kuiper Belt received top priority in an earlier survey, but the road to its success began many decades earlier!

In addition to the Decadal Survey's recommendations, a presidential administration may bring a campaign promise or an election mandate to the table, such as the Artemis Program to establish a Moon base for mining.

Even so, Mars remains at the center of the US exploration strategy with the goal of establishing a human outpost there in the near future. Of course, we have been hearing about such plans in one form or another since the 1970s, but perhaps this time… Returning humans to the Moon is a critical step towards that end, along with continuing investigations of martian conditions.

FIGURE 17.6 Doors to the NAS in Washington, DC, USA.

Over two decades of exploration has revealed that Mars was once a very different environment, pointing to a wet surface with a thicker atmosphere billions of years ago. NASA's next step in exploring the red planet is the Mars 2020 mission with its Perseverance rover and pint-sized Ingenuity helicopter. It launched on 30 July 2020 and arrived on 18 February 2021. Perseverance and Ingenuity are studying martian geology to assess its past habitability and signs of ancient life.

Perseverance is the latest in a long heritage of rovers deployed by NASA. It is as big as a mid-sized car, weighs about a ton, and will carry seven scientific instruments. The rover will gather rocks and soil samples for possible return to the Earth by a future mission. As the first helicopter, Ingenuity will primarily demonstrate this technology for future use. After completing system checks, it embarked on a 30-day campaign to explore its surroundings on 19 April 2021. As with almost all missions these days, Mars 2020 is an international collaboration with contributions from France, Spain, and Norway.

Since Americans last walked the Moon, China and India have sent robotic missions to our sister world. Nations increasingly consider the Moon a key strategic asset in outer space beyond scientific research. In effect, we are engaged in another international space race, the goal being to return humans to the lunar surface and even establish a colony.

In addition to its usefulness as a step towards a martian colony, returning to the Moon is itself another major NASA objective. The Artemis Program calls for building a sustainable outpost at the lunar south pole by 2024. Private companies will search for subsurface water for sustenance and for conversion to rocket fuel (Figure 17.7).

As part of the Artemis Program, NASA is taking the revolutionary approach of joining with commercial partners under the Commercial Lunar Payload Services program. NASA will provide heavy-lift vehicles, when needed, to carry commercial payloads and eventually people to the surface. Mining and other resource-exploitation activities are envisioned, including tourism. The companies will own the resources they mine.

This strategy raises a host of legal issues under existing space law. Yes, there really is such a thing, beginning with the Outer Space Treaty of 1967, which restricts the use of the Moon and other solar-system bodies to peaceful purposes. Following international agreements concerning Antarctica, the Treaty also states that the Moon is "not subject to national appropriation by claim of sovereignty, by means of use or occupation, or by any other means."

FIGURE 17.7   In 2020, the first complete geologic map of the Moon was released.

The US is drafting legal blueprints for lunar excavating as part of a new American-sponsored international agreement called the Artemis Accords, which will apply to all participants in the Artemis Program. In addition to NASA's traditional science, technology, and discovery roles, the proposed Accords require NASA to be a facilitator of American foreign policy, an unusual initiative at best.

Descriptions of future exploration plans from all space-faring nations of the world would fill several volumes, so we must limit our discussion to a few intriguing ones. They include extended operations of existing probes under NASA's slogan to "Reduce, Reuse, and Recycle," new primary missions of exploration [PM], the implementation of approved programs of solar-system exploration [IM], and the formulation of new mission concepts [FM].

The grand strategy for solar-system exploration is first reconnaissance (flyby for a brief snapshot), followed by orbiters (long-term behavior), then landers (detailed *in situ* surface and atmosphere studies), then sample return (investigations using Earth-based state-of-the-art technology), and finally human exploration. Mission designers must include planetary protection measures in all these efforts to prevent contamination of another world with lifeforms from the Earth and vice versa! Selected future missions are discussed next with letter designations introduced above to indicate their degree of development.

NASA's Europa Clipper (IM) fits into this scheme. It will return to the jovian system to orbit the Icy Satellite, Europa, and explore the hidden ocean beneath that moon's surface. Could life exist in this vast sea containing more water than all of Earth's oceans? The mission will not look for life itself, but rather learn more about the ocean, the icy geology, and the surface composition. Using results from this mission, NASA will plan a future lander that will search for life.

NASA's Lucy mission (IM) will make a grand tour of Jupiter's Trojan asteroids. Plans are to encounter seven Trojans, in both camps. Interesting in and of themselves, scientists think these primitive bodies hold clues for deciphering the history of the Solar System. NASA expects to launch Lucy in 2021 for a 12-year mission (Figure 17.8).

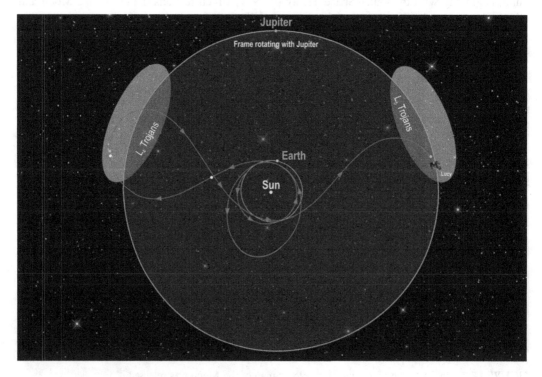

FIGURE 17.8 Anticipated Lucy mission to a Trojan.

In keeping with small-body exploration, NASA's Psyche[7] mission (IM) will travel to a unique metallic asteroid. Scientists suspect that this asteroid, also named Psyche, is the exposed iron-nickel core of an ancient body that was shattered by a collision in the Asteroid Belt. The orbiter will launch in 2022 with arrival in 2026. It will carry instruments to determine the composition of the asteroid, map the surface, characterize the interior structure, and measure its magnetic properties (Figure 17.9).

NASA's Double Asteroid Redirection Test [DART] is set to launch in 2021 to the binary NEO system, Didymos[8] and Dimorphos.[9] Using a kinetic impactor, this mission will perform the first asteroid trajectory deflection experiment, an important element of our planetary defense efforts. It will hit Dimorphos in 2022 to change its orbit slightly. ESA's Hera mission will follow to examine the impact results.

Another ambitious mission is NASA's Dragonfly (FM), designed to visit Saturn's largest moon, Titan, a high priority for planetary exploration. Recall that Titan is an ocean world with lakes of liquid methane and a dense atmosphere; it is home to methane clouds, rain, and snow. ➤ Chapter 12 ➤ This environment is an ideal destination for studying the conditions necessary for habitability and the chemical interactions that may have occurred before life developed on the Earth. Dragonfly is a helicopter drone that can fly from one site to another, exploring Titan's surface and atmosphere. NASA scheduled its launch for 2026, with arrival at Titan in 2034 (Figure 17.10).

The JUpiter ICy moons Explorer spacecraft ("JUICE"; IM) is the first flagship mission in ESA's Cosmic Vision program. It will launch in May 2022, arriving at the Jupiter system in 2029 to study Ganymede, Europa, and Callisto as well as Jupiter itself. These Icy Satellites are ocean worlds with subsurface global seas of water. Could life exist under the surface of these moons? The mission will address the conditions for habitability of ocean worlds and for the possible emergence of life.

At this time of writing, the BepiColombo (PM) mission is en route to Mercury, the least explored planet in the inner Planetary System; it should arrive there in 2025. One of ESA's cornerstone missions, BepiColombo will study and understand the composition, geophysics, atmosphere, magnetosphere, and history of Mercury. It consists of two individual orbiters: the

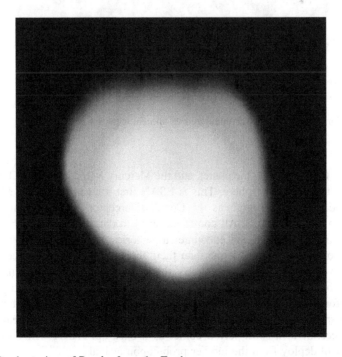

FIGURE 17.9 Our best view of Psyche from the Earth.

FIGURE 17.10   Surface of Titan as reconnoitered by the Huygens lander.

Mercury Planetary Orbiter to map the planet, and the Mercury Magnetospheric Orbiter supplied by JAXA to investigate its magnetosphere. This is ESA's first joint mission with JAXA.

Another joint ESA/JAXA program is the Comet Interceptor (FM) mission to a long-period comet preparing to launch in 2028. All comet missions to date have visited short-period comets since their orbits are well known and encounter trajectories can be computed easily. Remember that long-period comets are more pristine than their short-period cousins, since they may be on their initial trip to the inner Solar System. By the time one is discovered, insufficient time is available to scramble a mission. Therefore, the Comet Interceptor will park in a stable orbit after launch and wait for a new visitor to approach. An encounter with an interstellar comet is also a possibility. If the sponsors do not identify an appropriate target within a few years, then they will redirect their spacecraft to a short-period comet as a back-up. JAXA will contribute scientific instruments that will deploy from the mother probe upon arrival.

ExoMars 2022 (FM) is ESA's mission to return to Mars with a rover and surface platform provided by Russia. Russia will also provide a Proton rocket for launch and the descent module to the martian surface. ESA and Russia plan to launch in 2022 with the trip to Mars taking 9 months. The primary objective is landing the rover at a site with well-preserved organic material from the very early history of the planet. The rover will establish the physical and chemical properties of martian samples, mainly from the subsurface. At several sites, it will obtain underground samples using a special drill. These samples are more likely to include biomarkers, since the tenuous martian atmosphere offers little protection from ultraviolet radiation at the surface (Figure 17.11).

The Chinese have an ambitious space program for the rest of the 21st century. They are laying the groundwork to build a lunar research base for their taikonauts. The end of this decade may see China's space agency on Mars. Using experience gained from lunar exploration, Tianwen-1[10] [PM] was launched to Mars on 20 July 2020 with an orbiter and a rover (named Zhurong)[11] that will search for pockets of buried water using radar technology. Could life exist in these water pockets? It entered orbit around Mars in February 2021 and deployed Zhurong to the surface in May 2021. China has also announced plans for a Mars sample-return mission, a Jupiter orbiter, missions to Ice Giants, and interstellar space. Stay tuned for more details!

In addition to its international collaborations, Japanese programs under development include the Demonstration and Experiment of Space Technology for INterplanetary voYage Phaethon fLyby dUSt Science [DESTINY+] [FM] to the active asteroid Phaethon leaving in 2022. The Martian Moons eXploration [MMX] [FM] mission, set to launch in 2024, will return samples of its largest moon, Phobos, after observing Deimos and the martian atmosphere (Figures 17.12 and 17.13).

Having completed two missions to the Moon, India joined the ranks of space-faring nations. They plan to return to the Moon (Chandrayaan-3) [FM] and visit Mars (Mangalyaan-2) [FM], the Sun (Aditya-L1) [FM], Venus (Shukrayaan-1) [FM], and Jupiter. India is developing the ability to carry humans to space. They plan to send their vyomanauts first to a space station and later to the Moon.

The United Arab Emirates (UAE), among other countries, has joined the 'space club.' They launched a Mars orbiter on 19 July 2020, known as Hope [PM], with the assistance of Japan. The orbiter arrived at the red planet in early 2021 to study the martian atmosphere. The UAE Space Agency plans to send Rashid, a small rover, to the Moon in 2024.

Prior to the present century, space exploration was exclusively state sponsored. However, several recent government/private ventures indicate a trend towards increasingly commercialized endeavors. The goal is to share the enormous expense of space exploration. For example, NASA has selected three private companies to design craft carried by NASA's Orion rockets to land astronauts on the Moon as early as 2024. Amazon's Chief Executive Officer [CEO], Jeff Bezos, owns Blue Origin. Tesla's CEO Elon Musk owns SpaceX. Dynetics, owned by Leidos, is led by David King, a former NASA center director.

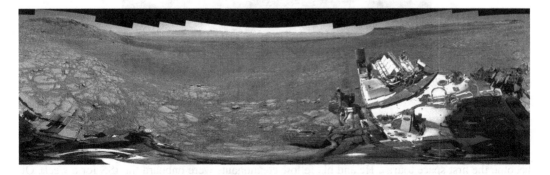

FIGURE 17.11 In 2020 NASA released this 360° panorama from Mars taken by the Curiosity Rover.

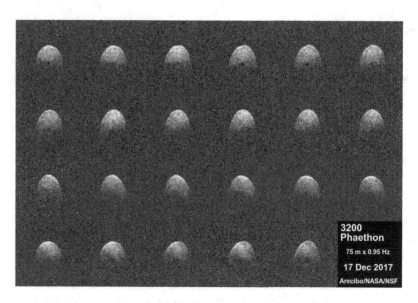

FIGURE 17.12 Best views of asteroid Phaethon from the Earth.

FIGURE 17.13 Phobos imaged by the Mars Reconnaissance Orbiter.

Interestingly, Russia pioneered the commercialization of space in the form of tourism, space travel for recreational purposes. In 2001, they sold a ticket on their Soyuz rocket that ferries people to the International Space Station [ISS]. American entrepreneur Dennis Tito spent $20 million to become the first space tourist. He and his fellow cosmonauts were onboard the ISS for a week. Of course, it was a round-trip ticket! Since this historic flight, seven additional people have made self-funded trips; the Hungarian-American Charles Simonyi even made two such journeys. Space tourists represent a variety of countries such as the US, South Africa, Iran, Hungary, and Canada.

Russia halted these jaunts in 2010, but similar excursions may resume if SpaceX and Boeing make good on their plans. In fact, NASA now seems to be willing to make the ISS available for commercial opportunities, including as a space hotel. It has contracted with the American company Axion to add luxury accommodations for the ultra-wealthy. The price of a 10-day space holiday would be $55 million, with an additional million for training prior to the trip. NASA also plans to allow the ISS to be used as a set to film major motion pictures. Perhaps NASA's role will transition to a travel agency in the not-to-distant future!

Marginally lighter on the pocketbook is a sub-orbital flight. The $10 million Ansari XPRIZE challenged a private company to reach an altitude of 100 kilometers twice within two weeks; this altitude is commonly considered the threshold for spaceflight. Iranian-American Anousheh Ansari, the third ISS excursionist and the first woman to tour it, funded the competition. An American company, Scaled Composites, won it in 2004 using their SpaceShipOne. They licensed their design to the British entrepreneur Richard Branson. His company Virgin Galactic modified the original design to create SpaceShipTwo, which holds six passengers and two pilots. Several hundred tickets have been sold at the original price of $200,000, which has now increased to $250,000. Everything seems to get more expensive these days!

The 2-hour itinerary for a SpaceShipTwo flight consists of launch from Spaceport America in New Mexico, a rapid accent to an altitude of 100 kilometers, a few minutes of weightlessness in space, and a near vertical descent returning to the Spaceport. Afterwards, Virgin Galactic awards passengers their astronaut wings for officially having traveled into space. If you like roller coasters, this is the ultimate thrill!

For two generations only, astronauts have been able to observe the Earth as a planet. American business magnate Paul Allen <1953–2018>, co-founder of Microsoft Corporation, was a pioneer in developing privately funded space travel through his company Stratolaunch Systems. Space tourism suggests the opportunity to allow others, coming from a variety of perspectives, to view the Earth from afar and interpret what they see. Undoubtedly, the most precious thing these new solar-system adventurers will encounter in space is the window.

The logical follow-up to orbital and suborbital rides would be lunar tourism. Google established the Lunar XPRIZE in 2007 with a purse of $30 million to spur affordable travel to the Moon. To win, a team must have landed on the Moon, traveled 500 meters, and established two-way communication with the Earth by 2018.

The Israeli company SpaceIL attempted to achieve the goals of the Lunar XPRIZE in 2019. As mentioned earlier, after traveling to the Moon, their Beresheet spacecraft and its tardigrade occupants unfortunately experienced what is called in the business, a 'hard landing.' As a consolation, the XPRIZE administrators gave SpaceIL a $1 million Moonshot Award for 'touching' the lunar surface. Undoubtedly, someone will achieve the goals originally set by the Google XPRIZE, opening the door for affordable travel to lunar hotels. Anyone up for hiking in a crater or for a round of lunar golf?

SpaceX is developing Starship, a private heavy-lift launch vehicle. It has at least one customer for a round-trip voyage around the Moon. In addition, it has plans to transport humans to Mars as soon as 2024 with this rocket.

Other private ventures to Mars have come and gone in recent years, such as one planning to send a married couple to the planet for a quick pass around it before returning home. Another would establish a small colony of experts on the martian surface—one-way travel where the participants presumably would become the first deaths on the red planet.

Life on Mars will not be easy. Water and oxygen will have to be mined. Perhaps shelters within lava tubes will protect visitors from solar radiation and provide some insulation from the surface cold. Plans to terraform the planet would follow. ➤ Chapter 8 ➤ Beyond Mars, how about a flight to Titan or into the wild black yonder?

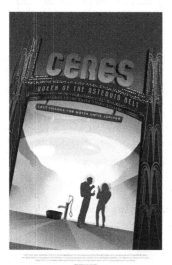

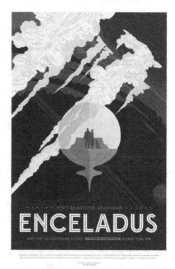

FIGURE 17.14   NASA 'tourist posters' (2018).

Another private initiative that is gaining attention is asteroid mining. A small asteroid could contain enough rare and precious minerals to make the venture cost effective. For instance, geologists estimate that a single, modestly sized asteroid might contain $50 billion worth of platinum. Concepts involve everything from completely robotic efforts to crewed missions of space miners. The metallic asteroid Psyche could be an interesting destination. The Psyche mission described above will provide more details for the prospective prospectors.

The future of space travel and exploration is bright. Already several private companies have rockets capable of lifting payloads to Earth orbit. This trend will continue as the century progresses, freeing NASA to concentrate on deep space travel. More countries will become spacefaring nations to explore other nooks and crannies in our Solar System. As you have seen, our Solar System contains no shortage of interesting destinations to visit, some even worlds of ice and fire! (Figure 17.14)

NOTES

1 Also known as exobiology.
2 Found in the Allan Hills region of Antarctica during the ANtarctic Search for METeorites [ANSMET] 1984–1985 season.
3 Alternatively called water bears or moss piglets.
4 Hebrew for 'in the beginning.'
5 The 16th US president, Abraham Lincoln ⟨1809–1865⟩, established this prestigious organization of distinguished scientists to provide the best independent and objective scientific advice for policy decisions.
6 Latest entitled, "Vision and Voyages for Planetary Science in the Decade 2013–2022."
7 Personification of the human soul in Greek and Roman mythology.
8 Greek for 'twin.'
9 Greek for 'having two forms.'
10 Meaning 'heavenly questions,' taken from a 4th century BCE poem.
11 The fire god in Chinese folklore.

Appendix I Units and Numbers

- We use metric units (with English units following in parentheses), following the Système Internationale (SI) standard. An exception is compound units, such as those for density, which are metric exclusively. To reduce the number of units with which the reader must be familiar, we restrict ourselves, when practical, to centimeters, meters, and kilometers (inches, feet, and miles) for length and grams and kilograms (tons) for mass. A metric prefix is commonly used that precedes a basic unit of measure indicating a multiple or fraction of the unit (see Table 1).
  - Units are written out, except for compound units: g = grams and cm = centimeters.
  - Be careful: Tons, not metric tonnes, are used. Tons are US tons (or "short tons," = 2,000 pounds avoirdupois).
  - The bar (used in the appendices only) may not be a familiar unit. It is roughly equal to the pressure of the atmosphere that you experience while standing on the surface of the Earth.

- Temperature is recorded in °C for Celsius (°F for Fahrenheit).
  - water boiling point = 100°C = 212°F
  - water freezing point = 0°C = 32°F
  - coldest temperature (absolute zero) = −273°C = −460°F

- Time is expressed in the familiar units of seconds, hours, days, weeks, months, and years. By international convention, times around the world are synchronized using Universal Time, Coordinated [UTC].

- Significant figures, also called significant digits, of a number are those digits that are meaningful within the precision of the measurement. This includes all digits except all leading zeros, trailing zeros that are merely placeholders, and spurious digits introduced by calculations carried out to greater precision than the original data (to be avoided).

Numbers in the text are cited to two significant figures when available. For example, 2 means between 1.5 and 2.5, while 2.0 means between 1.95 and 2.05.

- Scientific notation is used sparingly in the book for conciseness. It is a way of expressing numbers that are too big or too small to be conveniently written in decimal or fractional form. Astronomers, scientists, mathematicians, and engineers commonly use scientific notation because it can simplify certain mathematical operations. A number is written in scientific notation by using a decimal between 1 and 10 and multiplying it by a power of 10. For example, in chapter 15, we gave the power of the Sun as equivalent to four septillion 100-watt light bulbs. Written in scientific notation, this number is 4×10^{24}. The age of the Solar System, 4.6 billion years, can be written as 4.6×10^{9} years, or simply 4.6 Gyr.

Just how big are many numbers encountered in this book? Truly astronomical! Can we really appreciate how large they are? If you are lucky, you can hold about 100 individual objects in your mind at one time. Anything larger becomes an abstraction, merely a word or number with a lot of zeros, losing its real enormity. To appreciate large numbers, try this simple analogy with time: Count one number per second, so 100 becomes 100 seconds, or 1 minute, 40 seconds. Continuing to 1,000, you get 16 minutes, 40 seconds. So, the difference between 100 and 1,000 is 15 minutes. See where this is going? The next stop is 1,000,000. You will need 11 days, 13 hours, 46 minutes,

TABLE 1
Prefixes for Units from Large to Small

Prefix (symbol)	Spoken	Decimal notation	Scientific notation
Yotta (Y)	Septillion	1,000,000,000,000,000,000,000,000	10^{24}
Zetta (Z)	Sextillion	1,000,000,000,000,000,000,000	10^{21}
Exa (E)	Quintillion	1,000,000,000,000,000,000	10^{18}
Peta (P)	Quadrillion	1,000,000,000,000,000	10^{15}
Tera (T)	Trillion	1,000,000,000,000	10^{12}
Giga (G)	Billion	1,000,000,000	10^{9}
Mega (M)	Million	1,000,000	10^{6}
Kilo (k)	Thousand	1,000	10^{3}
Hecto (h)	Hundred	100	10^{2}
Deka (da)	Ten	10	10^{1}
	One	1	10^{0}
Deci (d)	Tenth	0.1	10^{-1}
Centi (c)	Hundredth	0.01	10^{-2}
Milli (m)	Thousandth	0.001	10^{-3}
Micro (μ)	Millionth	0.000001	10^{-6}
Nano (n)	Billionth	0.000000001	10^{-9}
Pico (p)	Trillionth	0.000000000001	10^{-12}
Femto (f)	Quadrillionth	0.000000000000001	10^{-15}
Atto (a)	Quintillionth	0.000000000000000001	10^{-18}
Zepto (z)	Sextillionth	0.000000000000000000001	10^{-21}
Yocto (y)	Septillionth	0.000000000000000000000001	10^{-24}

and 40 seconds (one million seconds). That is counting "24-7" with no resting allowed! So, the difference between 1,000 and 1,000,000 is more than a week. Reaching one billion? Would you believe you would need almost 32 years! The difference between a million and a billion is gigantic. We are fortunate if we live to see three billion seconds. There is one more stop to go. Counting to a trillion would take more time than recorded civilization: 31,710 years, 317 centuries, or over 3 millennia. One trillion is staggeringly colossal! So, when you read in Chapter 14 that there are more than a trillion comets in the Oort Cloud, we hope that you have a better appreciation of how many are thought to be there.

Dictionary of Units

Gram (g)—SI unit of mass used for measuring small quantities of a substance, 1 g = 0.035 ounces.
Kilogram (kg)—fundamental SI unit of mass roughly equivalent to 2 pounds; 1 kg = 1,000 g = 35 ounces = 2.2 pounds.
Megagram (Mg) = also known as the metric ton, or tonne, roughly equivalent to a ton; 1 Mg = 1,000 kg = 1,000,000 g = 1.1 ton = 2,205 pounds = 35,274 ounces.
Meter (m)—fundamental SI unit of distance roughly equivalent to 1 yard; 1 m = 39 inches.
Nanometer (nm)—SI unit of length used for extremely short distances; 1 nm = 1×10^{-9} m = 3.9×10^{-8} inches.
Millimeter (mm)—SI unit of length used for short distances; 1 mm = 10^{-3} m = 0.039 inches.
Kilometer (km)—SI unit of length used for planetary distances, a little more than half a mile; 1 km = 10^{3} m = 39,000 inches = 0.62 miles.

Metric ton (t)—megagram, or tonne.

Pascal (Pa)—metric base unit of pressure: at sea-level on Earth, we experience an atmospheric pressure (1 bar) of about 101,325 Pa (14.69595 pounds per square inch).

Megapascal (MPa)—metric unit of high pressure: at sea-level on Earth, we experience an atmospheric pressure (1 bar) of about 0.101325 MPa (14.69595 pounds per square inch); 1 MPa = 10^6 Pa = 150 pounds per square inch.

Système Internationale (SI)—the International System of Units, a worldwide standard system of measurement, commonly called the metric system.

Ton—customary unit of weight in the English system equal to 2,000 pounds; 1 ton = 2,000 pounds = 0.907185 Mg (metric tonne) = 907.185 kg = 32,000 ounces = 907,185 g.

Universal Time, Coordinated (UTC)—time kept by a world-wide network of atomic clocks that forms the basis of most civil time; each time zone in the United States is behind UTC by a specific number of hours.

Appendix II Properties of the Planets and Dwarf Planets

TABLE 1
Planets (and the Moon)

	Mercury	Venus	Earth	Moon	Mars	Jupiter	Saturn	Uranus	Neptune
Symbol	☿	♀	⊕	☾	♂	♃	♄	⛢	♆
Mass (10^{24} kg)	0.330	4.87	5.97	0.073	0.642	1898	568	86.8	102
Diameter (km)	4879	12,104	12,756	3475	6792	142,984	120,536	51,118	49,528
Density (g/cm^3)	5.427	5.243	5.514	3.340	3.933	1.326	0.687	1.271	1.638
Gravity (m/s^2)	3.7	8.9	9.8	1.6	3.7	23.1	9.0	8.7	11.0
Escape velocity (km/s)	4.3	10.4	11.2	2.4	5.0	59.5	35.5	21.3	23.5
Rotation period (hours)	1407.6	−5832.5	23.9	655.7	24.6	9.9	10.7	−17.2	16.1
Length of day (hours)	4222.6	2802.0	24.0	708.7	24.7	9.9	10.7	17.2	16.1
Distance from sun (au)	0.387	0.723	1.00	0.384*	1.524	5.204	9.582	19.201	30.047
Perihelion (au)	0.307	0.718	0.983	0.363*	1.381	4.950	9.041	18.324	29.709
Aphelion (au)	0.467	0.728	1.017	0.406*	1.666	5.459	10.124	20.078	30.385
Orbital period (years)	0.24	0.62	1.00	27.3d	1.88	11.86	29.42	83.75	163.72
Orbital velocity (km/s)	47.4	35.0	29.8	1.0	24.1	13.1	9.7	6.8	5.4
Orbital inclination (degrees)	7.0	3.4	0.0	5.1	1.9	1.3	2.5	0.8	1.8
Orbital eccentricity	0.205	0.007	0.017	0.055	0.094	0.049	0.057	0.046	0.011
Obliquity to orbit (degrees)	0.01	177.4	23.4	6.7	25.2	3.1	26.7	97.8	28.3
Mean temperature (°C)	167	464	15	−20	−65	−110	−140	−195	−200
Bond albedo	0.068	0.77	0.31	0.11	0.25	0.34	0.34	0.30	0.29
Surface pressure (bars)	0	92	1	0	0.01	Unknown#	Unknown#	Unknown#	Unknown#
Number of moons	0	0	1	0	2	79§	82§	27	14
Ring system?	No	No	No	No	No	Yes	Yes	Yes	Yes
Global Magnetic field?	Yes	No	Yes	No	No	Yes	Yes	Yes	Yes
	Mercury	Venus	Earth	Moon	Mars	Jupiter	Saturn	Uranus	Neptune

*The values for the Moon's orbit refer to its orbit about the Earth.
#The force of gravity for the Jovian Planets is at the visible cloud layer. Temperature also is listed for the visible cloud layer. These planets have no solid surface; their surface gas pressure is listed as "unknown."
§Jupiter has 78 confirmed and 1 provisional (unconfirmed) satellites; Saturn has 53 confirmed and 29 provisional ones.

TABLE 2
Dwarf Planets

	Ceres	Pluto	Haumea	Makemake	Eris
Symbol	⚳	♇	🝻	🝼	⯰
Mass (10^{21} kg)	0.939	13.0	4.01	3.1?	16.6?
Diameter (km)	950[1]	2374	1560[2]	1430	2326
Density (g/cm^3)	2.600	1.860	2.018	1.700?	2.520?
Gravity (m/s^2)	0.26	0.62	0.4	<0.57?	0.82?
Escape velocity (km/s)	0.50	1.2	0.81	<0.91?	1.38?
Rotation period (hours)	9.1	−153.3	3.9	22.5	25.9
Length of day (hours)	9.1	153.3	3.9	22.5	25.9
Distance from sun (au)	2.769	39.482	43.182	45.430	67.864
Perihelion (au)	2.559	30.164	34.766	38.104	38.272
Aphelion (au)	2.9780	48.494	51.598	52.756	97.457
Orbital period (years)	4.60	247.94	283.84	306.17	558.77
Orbital velocity (km/s)	17.9	4.7	4.5	4.4	3.4
Orbital inclination (degrees)	10.6	17.2	28.2	29.0	44.0
Orbital eccentricity	0.076	0.249	0.195	0.161	0.436
Obliquity to orbit (degrees)	4	122.5	13.8	46–78	78?
Mean temperature (°C)	−105	−225	−235**	−241**	−245**
Bond albedo	0.034	0.72	0.33	0.62	0.55
Surface pressure (bars)	0*	0.00001§	Unknown#	Unknown#	Unknown#
Number of moons	0	5	2	1^	1
Ring system?	No	No	Yes	Unknown	Unknown
Global magnetic field?	No	No	Unknown	Unknown	Unknown
	Ceres	Pluto	Haumea	Makemake	Eris

Source: Table I by NASA; Table II by NASA and current scientific literature.
* Trace amounts of water vapor have been detected.
§ Has a transient atmosphere around perihelion.
# May have a transient atmosphere around perihelion, like Pluto.
^ Unconfirmed discovery.
** Equilibrium surface temperature at mean distance from the Sun.

NOTES

1 Tri-axial ellipsoid with axes: 965 km × 961 km × 891 km.
2 Tri-axial ellipsoid with axes: 2100 km × 1680 km × 1074 km.

Appendix III Gazetteer of Solar-system Geographical Features

Gazetteer of solar-system geographical features

Name	Landform type	Location	Latitude	Longitude
Aeolis/Gale	Mountain/Crater	Mars	−5	138
Aiken	Basin	Moon	−16	173
Allan Hills	Range	Earth	−77	160
Barringer	Crater	Earth	35	−111
Belgica	Scarp	Mercury	−50	−64
Caloris	Basin	Mercury	31	−162
Chicxulub	Crater	Earth	21	−90
Cleopatra	Crater	Venus	66	7
Côlonia	Crater	Earth	24	−47
Copernicus	Crater	Moon	10	−10
Enterprise	Scarp	Mercury	−37	−77
Everest	Mountain	Earth	28	87
Gipul	Chain	Callisto	70	−50
Grand Canyon	Valley	Earth	36	−112
Great Red Spot	Storm	Jupiter	−22	NA
Hellas	Plain	Mars	−42	71
Herschel	Crater	Mimas	3	110
Himalayas	Range	Earth	29	−84
Kaua'i	Volcano	Earth	22	−159
Kilauea	Volcano	Earth	19	−155
Korolev	Plain	Moon	−4	−157
Lö'ihi	Volcano	Earth	19	−155
Manicouagan	Crater	Earth	51	−69
Mare Crisum	Basin	Moon	16	59
Mariana	Trench	Earth	11	−142
Marineris	Valley	Mars	−14	−59
Maui	Volcano	Earth	21	−156
Maxwell	Mountain	Venus	65	3
Nördlingen	Crater	Earth	49	11
Odysseus	Crater	Tethys	33	−129
Olympus Mons	Volcano	Mars	19	−134
Rachmaninoff	Basin	Mercury	28	−57
Saint Helens	Volcano	Earth	46	−122
Sputnik	Plain	Pluto	25	175
Tharsis	Plain	Mars	0	−100
Tooting	Crater	Mars	23	−152
Vredefort	Crater	Earth	−27	28

Craters on the Moon are named for famous astronomers.

Craters on Mercury are named for artists, composers, or poets.

Craters on Mars are named for famous scientists and science-fiction writers.

Mountains (Montes) on Venus are usually named for goddesses.

Scarps (Rupes) on Mercury are named after ships of discovery.

[Latitude and longitude are measured in degrees. Longitude increases to the East.]

Appendix IV Planetary and Related Space Missions

(Font codes: Normal—completed; **Bold**—ongoing; *Italics*—in development or on its way).
  (Letter codes: L—launch date, S—arrival or start of mission, E—end of mission).
  (Date format: month-day-year).

Aditya-L1 (ISRO)—mission to study the Sun's photosphere, chromosphere, and corona from the L1 Lagrange point. L:2022. Sanskrit (आदित्य) for Sun (Chapter 17).

Advanced Composition Explorer [ACE] (NASA)—spacecraft positioned at a Lagrange point to detect high-energy particles from the Sun and other sources. L:08-25-1997, S:01-21-1998 (Chapter 15).

Akatsuki (JAXA)—aka Venus Climate Observer [VCO], mission to study the three-dimensional motions and meteorology of the atmosphere of Venus. Failed to enter orbit in 2010 but was successfully inserted into an alternate orbit in 2015. L:05-20-2010, S:12-07-2015 (Chapter 6).

Apollo (NASA)—series of 16 manned missions to the Moon, each carrying three astronauts. Six missions successfully landed two astronauts on the lunar surface and returned them to the Earth; A11 (L:07-16-1969, S:07-20-1969), A12 (L:11-14-1969, S:11-18-1969), A14 (L:01-31-1971, S:02-05-1971), A15 (L:07-16-1971, S:07-29-1971), A16 (L:04-16-1972, S:04-20-1972), and A17 (L:12-07-1972, S:12-10-1972). A13 (L:04-11-1970, S:04-14-1970) experienced an explosion *en route* to the Moon and returned to the Earth without landing (Introduction, and Chapters 4, 7).

BepiColumbo (ESA/JAXA)—mission to Mercury consisting of two orbiters riding together until arrival: the Mercury Planetary Orbiter [MPO] (ESA) and Mercury Magnetospheric Orbiter [MMO] (JAXA). Expected to flyby Venus twice *en route* to Mercury (2020). Six flybys of Mercury until orbital insertion (2021-2025). L:10-20-2018, S:10-02-2021 (1st flyby), 12-05-2025 (orbit), E:2026 (Chapter 5).

FIGURE A-1   Assembling the BepiColumbo spacecraft in an ESA clean room.

Beresheet (SpaceIL)—private Israeli Moon mission with lander; crashed into the lunar surface. L:02-22-2019, S:04-04-2019, E:04-11-2019 (Chapter 17).

Cassini-Huygens (NASA/ESA/ASI)—orbiter to the Saturn System with atmospheric Huygens Probe to Titan (01-14-2005), Grand Finale sent into Saturn's upper atmosphere and burned up to prevent risk of contamination. On its way to Saturn, Cassini underwent gravity assists from two Venus flybys, one lunar flyby, and one Jupiter flyby, in addition to a flyby of asteroid 2685 Masursky (S-type). L:10-15-1997, S:07-01-2004, E:09-15-2017 (Introduction and Chapters 10, 12).

FIGURE A-2   Assembly of the Cassini-Huygens spacecraft to Saturn.

Chandrayaan-1 (ISRO)—orbiter with probe deployed to the Moon to map its surface mineralogy; impact probe landed near the lunar south pole. Sanskrit (**चन्द्रयान**) for 'Moon craft.' L:10-22-2008, S:11-08-2008, E:08-28-2009 (Chapter 4).

Chandrayaan-3 (ISRO)—mission to land and explore the south pole of the Moon with a rover. The successor to Chandrayaan-2, that was lost when attempting to land on the lunar surface in 2019, will use the orbiter from the previous mission. Sanskrit (**चन्द्रयान**) for 'Moon craft.' L:2022 (Chapter 17).

Chang'e 1 (嫦娥一號) (CNSA)—first Chinese orbiter to the Moon and deorbited to lunar surface at end of mission. L:10-24-2007, S:11-05-2007, E:03-01-2009 (Chapter 4).

Chang'e 4 (嫦娥一四) (CNSA)—Chinese orbiter and Yutu 2 rover that landed near the lunar south pole on 01-03-2019; first mission to the lunar farside. L:12-07-2018, S:12-12-2018 (Chapter 4).

FIGURE A-3   Chang'e 4 (left) with Yutu 2 rover (right) on the lunar farside.

Chang'e 5 (嫦娥一號) (CNSA)—China's first lunar sample-return mission. Landed in a young volcanic region in Oceanus Procellarum, drilled about 1 meter (3 feet) deep to collect almost 2 kilograms (4 pounds) of lunar rocks, and returned them to Earth. L:11-23-20, S:12-01-20 (lander), E:12-16-20 (Chapter 4).

CHaracterising ExOPlanets Satellite [CHEOPS] (ESA)—space telescope to discover exoplanets using the Transit Method. L:12-18-2019, S:02-07-2020 (Chapter 16).

Clementine (Ballistic Missile Defense Organization/NASA)—polar orbiter to the Moon that discovered ice at the lunar poles. L:01-25-1994, S:02-19-1994, E:08-08-1994 (Chapter 4).

Comet Interceptor (ESA/JAXA)—spacecraft with two small probes to wait at the L2 Lagrange point for an encounter with long-period comet. L:2028 (Chapter 17).

Convection, Rotation and planetary Transits [CoRoT] (France)—small space telescope for detecting exoplanets using the Transit Method. Instrument ceased operations in November 2012. Deorbited in June 2014. L:12-27-2006, S:01-18-2007, E:03-31-2013 (Chapter 16).

COsmic Background Explorer [COBE] (NASA)—Earth satellite, first to map the cosmic background radiation (CBR) from the Big Bang; a Nobel prize was awarded to two members of the COBE team for discovering the anisotropy of the CBR. L:11-18-1989, E:12-23-1993 (Chapter 1).

Curiosity (NASA)—martian rover to study habitability, part of the **Mars Science Laboratory** [MSL]. L:11-26-2011, S:08-06-2012 (Chapter 8).

Dawn (NASA)—first mission to explore a dwarf planet, orbited Vesta (July 2011–September 2012) and Ceres (March 2015–June 2017) using solar-ion propulsion. L:09-27-2007, S:09-27-2007, S:07-16-2011 (Vesta), S:03-06-2015 (Ceres), E:11-01-2018 (Chapter 9).

Deep Impact (NASA) - space probe to flyby Comet 9P/Tempel 1 with impactor to reveal subsurface material. Extended mission EPOXI. L:01-12-2005, S:07-04-2005, E:11-04-2010 (Chapter 13).

Deep Impact–EPOXI (NASA)—Extrasolar Planet Observation and Deep Impact Extended. Flyby investigation [DIXI] to Comet 103P/Hartley 2. S:11-04-2010, E:08-08-2013 (Chapter 13).

Deep Space 1 (NASA)—space probe to Comet 19P/Borrelly, primarily to test twelve, new, revolutionary technologies, including solar-ion propulsion engines. Flyby of asteroid Braille (07-29-1999) (Q-type). L:10-24-1998, S:09-22-2001, E:12-18-2001 (Chapter 13).

Demonstration and Experiment of Space Technology for INterplanetary voYage Phaethon fLyby dUSt science [DESTINY+] (JAXA) - space probe flyby of Phaethon, the parent body of the Geminid meteor shower, and *in situ* analysis of interplanetary dust. L:2024, S:2028, E:2030 (Chapter 17).

Double Asteroid Redirection Test [DART] (NASA)—spaced probe to crash into potentially hazardous, near-Earth binary asteroids in order to deflect them and test planetary protection plans. If successful, ESA's Hera mission will observe its aftermath in 2026. L:2021, S:2022, E:2022 (Chapter 17).

Dragonfly (NASA)—orbiter and lander with helicopter drone to Saturn's moon, Titan, to investigate Titan's atmosphere, surface, subsurface, and origins-of-life issues. L:2026, S:2034 (Chapter 17).

Europa Clipper (NASA)—orbiter to Jupiter System to investigate whether ocean-world Europa has conditions suitable to harbor life. L:2024 (Chapter 17).

ExoMars (ESA/Roscosmos)—lander and rover to search for signs of life on Mars. L:2022 (Chapter 17).

Explorer (US Army/NASA)—series of small-scale missions launched first by the US Army Ballistic Missile Agency and, then, NASA, after its founding; Explorer 1 (01-31-1958) and 3 (03-26-1958) were the first successful American satellites; detected and confirmed the Van Allen Radiation Belts. L:01-31-1958, E: 08-24-1958 (Chapter 7).

FIGURE A-4   Explorer 1 full-scale model with Jupiter-C first stage model rocket.

Gaia (ESA)—Earth-orbiting telescope at the L2 Lagrange point used to make precise maps of stellar positions and motions of star in the Milky Way Galaxy; as a secondary mission, discovered numerous small bodies in the Solar System. L:12-19-2013, S:01-08-2014, E:12-31-2022 (Chapter 16).

Galileo (NASA)—orbiter to Jupiter System with probe into its atmosphere; launched from Space Shuttle Atlantis; sent into Jupiter's upper atmosphere at end of mission to burn up and prevent risk of contamination; asteroid flybys: 951 Gaspra (10-29-1991) (S-type) and 243 Ida (S-type) and Dactyl (08-28-1993); space observations of Comet D/Shoemaker-Levy 9 collision with Jupiter (07-22-1994). L: 10-18-1989, S:12-07-1995, E:09-21-2003 (Chapters 9, 10, 12).

Giotto (ESA) - space probe to comets 1P/Halley (03-13-1986) and to 26P/Grigg-Skjellerup (07-10-1992) first close-up images of a comet's nucleus. L: 07-02-1985, S:03-13-1986, E:07-23-1992 (Chapter 13).

Gravity Recovery And Interior Laboratory [GRAIL] (NASA) - twin orbiters, Ebb and Flow, sent to the Moon in order to probe the lunar interior by measuring gravitational effects; intentionally crashed into the Moon to probe subsurface material; observed by Lunar Reconnaissance Orbiter. L: 09-10-2011, S:01-01-2012, E:12-17-2012 (Chapter 4).

Hayabusa 1 (JAXA)—mission to rendezvous with (09-01-2005), land on (11-19-2005), and returned a sample (06-13-2010) of the small near-Earth asteroid Itakawa (S-type); used solar-electric propulsion. L:05-09-2003, S:09-01-2005, E:06-13-2010 (Chapter 9).

Hayabusa 2 (JAXA)—mission to the small near-Earth asteroid Ryugu (C-type) to: rendezvous (06-27-2018); deploy four rovers, including the Mobile Asteroid Surface Scout [MASCOT] (DLR/CNES) (10-03-2018); impact the surface (04-05-2019); collect two subsurface samples (07-11-2019); and return the samples to the Earth (12-05-2020); extended mission will visit another asteroid. L:12-04-2014, S:06-27-2018, E:2031 (Chapter 9).

FIGURE A-5   Hayabusa 2 with MASCOT rover ready to go to asteroid Ryugu.

Helios (Germany/NASA)—pair of orbiters, Helios 1 and Helios 2, sent to the Sun within Mercury's orbit; Helios 2 reached perihelion of 0.29 au (04-17-1976), a record at that time; observed tails of three comets. L:12-10-1974 (1), 01-15-1976 (2), S:01-16-1975 (1), 07-21-1976, E:02-18-1985 (1), 12-23-1979 (2) (Chapter 15).

Hera (ESA)—mission to inspect the aftermath of DART's collision with a binary asteroid to access deflection planning results. L:2024, S:2026 (Chapter 17).

Hinode (JAXA/NASA/ESA)—mission to study the Sun's magnetic activity, cycle, eruptive phenomena, flares, and coronal mass ejections. L:09-23-2006, S:10-28-2006, E:2022 (Chapter 15).

Hiten (JAXA)—mission to test technologies for future lunar and planetary missions. L:01-24-1990, S:03-19-1990, E:04-11-1993 (Chapter 4).

Hope (الأمل) [Emirates Mars Mission—EMM] (UAESA/JAXA)—orbiter to Mars in order to study its atmosphere and interaction with the solar wind. L:07-19-2020, S:02-09-2021, E:2023 (Chapter 17).

Hubble Space Telescope [HST] (NASA)—one of NASA's Great Observatories for imaging the Universe in visible and ultraviolet light; deployed with faulty optics that were corrected during a Space Shuttle mission; serviced five times by Space Shuttle crews; workhorse of astronomical observations that ushered in a new era of astronomy. L:04-24-1990, S:05-20-1990 (Chapter 16).

Ingenuity (NASA)—see **Mars 2020 Perseverance Rover.**

Interior Exploration using Seismic Investigations, Geodesy and Heat Transport [InSight] (NASA) –lander to Mars in order to study the geophysics of its deep interior. L:05-05-2018, S:11-26-2018 (Chapter 8).

Interplanetary Kite-craft Accelerated by Radiation Of the Sun [IKAROS] (JAXA)—first spacecraft to test solar-sail technology in interplanetary space; launched with **Akatsuki**, which went to Venus. L:05-21-2010, S:06-11-2010 (Chapter 6).

Interstellar Boundary EXplorer [IBEX] (NASA)—satellite to study energetic particles from the heliopause, the boundary at which the solar wind encounters the interstellar medium; launched using a Pegasus rocket from an airplane. L:10-19-2008, S:02-01-2009 (Chapter 15).

James Webb Space Telescope [JWST] (NASA/ESA/CSA)—replacement space telescope for the **Hubble Space Telescope** with about ten times the light gathering power; will operate in visible and near infrared light to study the Universe, with time allocated to Solar System studies. L:2021 (Chapter 16).

FIGURE A-6   Construction of the James Webb Space Telescope in a clean room.

Juno (NASA)—orbiter to the Jupiter System to probe its interior and atmosphere in polar orbit. L:08-05-11, S:07-04-16 (Chapter 10).

JUpiter ICy moons Explorer [JUICE] (ESA/NASA)—orbiter to the Jupiter System to make detailed observations of Jupiter and its satellites, Ganymede, Callisto and Europa. L:2022, S:2029, E:2032 (Chapter 17).

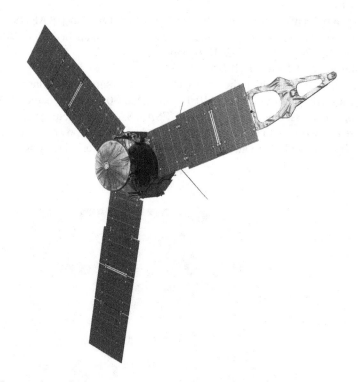

FIGURE A-7   Juno probe to Jupiter.

Kaguya (JAXA)—see SELenological and ENgineering Explorer.

Kepler Space Telescope (NASA)—space telescope for detecting exoplanets using the Transit Method; most productive such satellite with over 3,000 discoveries; when its coolant was exhausted, an extended mission was approved called K2. L:03-07-2009, S:05-13-2009, E:10-30-2018 (Chapter 16).

Kuiper Airborne Observatory [KAO] (NASA)—C-141A aircraft with mounted telescope to observe high in Earth's atmosphere, above most of the water vapor, and make infrared observations; replaced by **Stratospheric Observatory for Infrared Astronomy** [SOFIA] in 2010. S:05-21-1975, E:09-27-1995 (Chapter 11).

Lucy (NASA)—spacecraft to perform a grand tour of seven Trojan asteroids of Jupiter, five at the L4 Lagrange point (2027-2028) and two at L5 (2033), and a main-belt asteroid Donaldjohanson (2025) (C-type). L:2021, S:2027, E:2034 (Chapter 17).

Luna (USSR)—long series of Soviet spacecraft missions to the Moon between 1959-1976; Luna 3 first orbited Moon and made the first images of its farside (1959) (Chapter 4).

Lunar Orbiter (NASA)—series of orbiters to the Moon to map its surface for suitable Apollo landing sites; they were crashed into the Moon to prevent interference with any future spacecraft. L:1966-1967, S:1966-1968, E: 10-29-1966 until 01-31-1968 (Chapter 4).

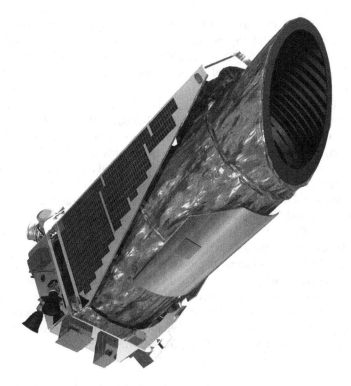

FIGURE A-8   Kepler space telescope, hunter of exoplanets.

Lunar Prospector (NASA)—orbiter to the Moon to map its surface composition; found ice at both poles; controlled impact on the lunar surface that was studied from Earth. L:01-07-1998, S:01-12-1998, E:07-31-1999 (Chapter 4).

Lunar Reconnaissance Orbiter [LRO] (NASA)—orbiter to the Moon to map its surface in great detail for planning future lunar missions; launched with Lunar CRater Observation and Sensing Satellite [LCROSS], which impacted the Moon (10-09-2009). L:06-18-2009, S:06-22-2009 (Chapter 4).

Magellan (NASA)—orbiter to Venus to map its surface using radio ranging techniques and study its atmosphere; launched from the Space Shuttle; made a controlled entry into Venus's atmosphere and burned up. L:05-04-1989, S:08-10-1990, E:10-13-1994 (Chapter 6).

Mangalyaan-2 (ISRO)—also known as Mars Orbiter Mission 2 [MOM 2], an orbiter to Mars with possible lander and rover; follow-up to **Mangalyaan-1**, India's first orbiter to Mars in 2014; Sanskrit (**मंगलयान**) for 'Mars craft.' L:2025 (Chapter 17).

Mariner (NASA)—series of ten missions launched between 1962 and 1973; Mariner 2, 5, and 10 flew by Venus (1962-1973), executing the first gravity assist maneuver; 10 also flew by Mercury (1974-1975); 4 took the first images of Mars (S:07-14/15-1965); 6 (S:07-31-1969) and 7 (S:08-05-1969) flew by Mars and returned an images of Phobos; 9 was the first spacecraft to orbit another planet (1971-1972) and imaged Deimos; 1, 3, and 8 were lost at launch (Chapters 5, 6, 8, 9).

Mars Atmosphere and Volatile Evolution Mission [MAVEN] (NASA)—Mars orbiter to study the martian atmosphere and its interaction with the solar wind; results indicate that such interactions may have caused the loss of the martian atmosphere over time. L:11-18-2013, S:09-21-2014 (Chapter 8).

Mars Exploration Rovers—Opportunity and Spirit (NASA) –twin rovers to Mars that found that the planet was awash in water in the ancient past, conditions that possibly could have supported microbial life. L:06-10-2003 (Spirit), 07-08-2003 (Opportunity), S:01-04-2004 (Spirit), 01-25-2004 (Opportunity), E: 05-25-2011 (Spirit), 02-13-2019 (Opportunity) (Chapter 8).

Mars Global Surveyor [MGS] (NASA)—mapping orbiter to study the martian surface, including changes in gullies, meteorology of the atmosphere, magnetic field, and interior. L:11-07-1996, S:09-12-1997, E:11-14-2006 (Chapter 8).

Mars Odyssey (NASA)—orbiter to Mars to study its climate, surface mineralogy, and water content and to prepare for human exploration. L:04-07-2001, S:10-24-2001 (Chapter 8).

Mars Pathfinder (NASA)—lander to Mars, named the Carl Sagan Memorial Station, and the first rover, Sojourner, named after African-American civil rights crusader, Sojourner Truth; among other findings, frequent dust devils were seen; successfully tested airbag landing technology. L:12-04-1996, S:07-04-1997, E:09-27-1997 (Chapter 8).

Mars Reconnaissance Orbiter [MRO] (NASA)—Mars orbiter imaging the planet for several martian years in order to study its climate and identify surface features associated with water. L:08-12-2005, S:03-10-2006 (Chapter 8).

Mars 2020 Perseverance Rover (NASA)—rover mission to Mars with the helicopter, **Ingenuity**, to search for signs of habitable conditions on Mars in the ancient past and past microbial life. A martian meteorite was returned to Mars for instrument calibration. L:07-30-2020, S:02-18-2021 (rover), 04-19-2021 (helicopter), E:2023, (Chapter 8).

Mars Science Laboratory [MSL]—Mars lander, better known as **Curiosity**, to evaluate the chemical make-up of surface features and the radiation environment in order to determine the likelihood of life. L:11-26-2011, S:08-06-2012 (Chapter 8).

FIGURE A-9 Curiosity rover exploring Mars.

Martian Moons eXploration [MMX] (JAXA)—mission to return samples of Mars's largest moon, Phobos; observe Deimos and the martian atmosphere. L:2024, S:2025, E:2029 (Chapter 17).

MErcury Surface, Space ENviroment, GEochemistry, and Ranging [MESSENGER] (NASA)—spacecraft to study Mercury; it flew by the planet three times in 2008–2009, then orbited the planet from 2011 to 2015, and finally ended its mission with a deliberate crash into the surface; en route to Mercury, it flew by Venus twice in 2006–2007. L:08-03-2004, S:03-17-2011, E:04-30-2015 (Chapter 5).

Near Earth Asteroid Rendezvous [NEAR-Shoemaker] (NASA)—orbiter to study near-Earth asteroid Eros (S-type); first to orbit and land on an asteroid; flyby of asteroid Mathilde (06-27-1997) (C-type). L:02-17-96, S:04-30-2000, E:02-28-2001 (Chapter 9).

New Horizons (NASA)—space probe to explore the Pluto-Charon System & a Kuiper Belt Object (Arrokoth), including flybys of asteroid 132524 APL (2006) (S-type) and Jupiter for a gravity assist (2007). L:01-19-06, S:07-14-15 (Pluto), S:01-01-19 (Arrokoth) (Chapter 14).

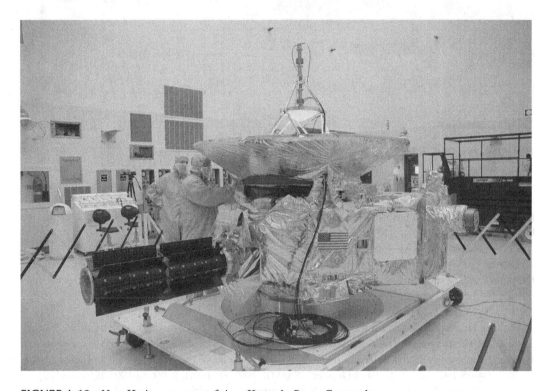

FIGURE A-10   New Horizons spacecraft in a Kennedy Space Center clean room.

Opportunity (NASA)—see Mars Exploration Rovers—Opportunity and Spirit.

Origins, Spectral Interpretation, Resource Identification, Security-Regolith Explorer [OSIRIS-Rex] (NASA)—space probe to potentially hazardous asteroid Bennu (C-type) to study its surface, collect a sample (10-20-2020) and return it to Earth in 2023; it also will measure the Yarkovsky effect. L:09-08-2016, S:12-03-2018, E:09-24-2023 (Chapter 9).

FIGURE A-11   OSIRIS-REx alongside its payload fairing.

Parker Solar Probe [PSP] (NASA)—orbiter to observe the Sun, its corona, and the solar wind; a collaborative mission with **Solar Orbiter**; will execute several Venus flybys (2018-2024) and pass through the corona within 6.2 million kilometers of the Sun's photosphere. L:08-12-18, S:10-03-2018, E:2025 (Chapter 15).

Perseverance (NASA)—see **Mars 2020 Perseverance Rover.**

Pioneer 10 (NASA)—space probe that executed the first flyby of the Jupiter System, imaging the planet and its moons, and measuring its magnetosphere; sibling probe to Pioneer 11; presently outside the heliosphere, on course to leave the Solar System in about 10,000 years. L:03-02-1972, S:12-03-1973, E:01-23-2003 (Chapter 10).

Pioneer 11 (NASA)—second space probe flyby of the Jupiter System and first flyby of the Saturn System (1979), imaging the planet and its moons, and measuring its magnetosphere; sibling probe to Pioneer 10; presently in the outer heliosphere, on course to leave the Solar System in thousands of years. L:04-05-1973, S:12-02-1974 (Jupiter), S:09-01-1979 (Saturn), E:09-30-1995 (Chapter 10).

Pioneer Venus (NASA)—two spacecraft, an orbiter and a multiprobe, sent to Venus; the orbiter carried an assortment of instruments for investigating plasma in the upper venusian atmosphere, observed reflected sunlight from the cloud layers at a variety of wavelengths, and, in addition carried a surface radar mapper; deployed atmospheric entry probes that sent data as they descended. L:05-20-1978, S:12-04-1978, E:08-08-1992 (Chapter 6).

Psyche (NASA)—after a Mars flyby (2023), orbiter to the metal-rich asteroid Psyche (M-type). L:2022, S:2026 (Chapter 17).

Rashid [Emirates Lunar Mission—ELM] (UAESA)—small rover to the Moon in order to study its surface composition and dust properties as they interact with the solar wind. L:2024 (Chapter 17).

Roman Space Telescope [RST] (NASA)—space telescope to observe the Universe in infrared light; formerly known as the Wide Angle Infrared Space Telescope [WFIRST] and replacement for the Spitzer Space Telescope [SST]; named after Nancy Grace Roman, NASA's first chief astronomer. L:2025 (Chapter 16).

Rosetta-Philae (ESA)—mission to rendezvous with Comet 67P/Churyumov-Gerasimenko in order to study its nucleus and environment for nearly two years; Philae was the first probe to land on a cometary surface (11-12-2014); Rosetta also imaged two asteroids, Šteins (09-05-2008) (E-type) and Lutetia (07-10-2010). L:03-02-2004, S:08-06-2014, E:09-30-2016 (Chapter 13).

Sakigake (Japan)—executing a flyby of Comet 1P/Halley, first Japanese deep space probe; measured comet-solar wind interaction on the sunward side. L:01-08-1985, S:03-11-1986, E:11-15-1995 (Chapter 13).

SELenological and ENgineering Explorer [SELENE] (JAXA)—also known as Kaguya, Japan's second lunar orbiter; performed a global survey of the composition of the lunar surface and ended with a controlled impact on the nearside. L:09-14-2007, S:10-03-2007, E:06-10-2009 (Chapter 4).

Shukrayaan-1 (ISRO)—orbiter with atmospheric balloon to study the surface and atmosphere of Venus; 'Venus craft' in Sanskrit (**शुक्रयान**). L:2024 (Chapter 17).

Sojourner (NASA)—see Mars Pathfinder.

Solar and Heliospheric Observatory [SOHO] (ESA/NASA)—positioned at Earth's L1 point to study the Sun from core to corona; observed thousands of sun-grazer comets. L:12-02-1995, S:05-01-1996 (Chapter 15).

Solar Dynamics Observatory [SDO] (NASA)—geosynchronous satellite to study the Sun's dynamic activity and space weather effects on the Earth. L:02-11-2010, S:05-14-2010, E:2030 (Chapter 15).

Solar Orbiter [SolO] (ESA/NASA)—with the Parker Space Probe, determine where the solar wind with its magnetic field originates, how its violent eruptions are produced, and how the Sun generates its magnetic field. L:02-10-2020, S:11-01-2021, E:2027 (Chapter 15).

FIGURE A-12 Rosetta spacecraft being prepared to chase a comet.

FIGURE A-13   SOHO satellite.

Solar TErrestrial Relations Observatories [STEREO] (NASA)—twin spacecraft positioned in orbit ahead of and behind the Earth in its orbit to create stereo images of the Sun; observed a disconnection event in Comet 2P/Encke (2007) and sun-grazer C/2011 W3 (Lovejoy) (12-14-2011). L:10-25-2006, S:2006, E:10-01-2014 (STEREO-B) (Chapter 15).

Spirit (NASA)—see Mars Exploration Rovers—Opportunity and Spirit.

Spitzer Space Telescope [SST] (NASA)—one of NASA's Great Observatories to observe the Universe in infrared light, with time dedicated for Solar System observations; its coolant was depleted in 2009 and it began warm operations; to be replaced by the *Roman Space Telescope*. L:08-25-2003, S:08-30-2003, E:01-30-2020 (Chapter 16).

Stardust (NASA)—mission to collect and return dust from Comet 81P/Wild 2 to the Earth (01-15-2006); imaged the comet's nucleus and environment; flyby of asteroid Annefrank (11-02-2004) (S-type). L:02-07-1999, S:01-02-2004, E:03-24-2011 (Chapter 13).

Stardust-NExT (NASA)—New Exploration of Tempel 1, Stardust's extended mission to Comet 9P/Tempel 1; imaged its nucleus and environment. S:02-15-2011, E:03-24-2011 (Chapter 13).

Suisei (Japan)—executing a flyby of Comet 1P/Halley, second Japanese deep space probe; imaged the comet's hydrogen envelope. L:08-18-1985, S:03-08-1986, E:02-22-1991 (Chapter 13).

Surveyor (NASA)—series of seven robotic landers to Moon from 1966-1968; intended to pave the way for landing humans on the Moon as part of the Apollo lunar landings (Chapter 4).

Tianwen-1 (天問) [TW-1] (CNSA)—Chinese orbiter with rover (named Zhurong) mission to study Mars's soil characteristics and distribution of surface water ice, surface material composition, atmospheric ionosphere and climate, and internal structure. L:07-22-2020, S:02-10-2021 (lander), 05-14-2021 (rover), E:2023 (Chapter 17).

Transiting Exoplanet Survey Satellite [TESS] (NASA)—satellite to search for exoplanets, including those that could support life, using the Transit Method. TESS will survey 200,000 of the brightest stars near the Sun to search for transiting exoplanets. L:04-18-2018, S:08-07-2018 (Chapter 16).

Ulysses (NASA/ESA/CSA)—deployed by the Space Shuttle to Jupiter to boost the spacecraft into a high inclination solar orbit and sent to the Sun to study its polar regions and the solar wind at high latitude; the Comet Watch Network (1992) imaged many comets: Borrelly, D'Arrest, Encke, Mueller, Pons-Winnecke, Temple 2, Tuttle, Hale-Bopp, and others. L:10-06-1990, S:06-26-1994, E:06-30-2009 (Chapter 15).

Vega 1 & 2 (ВеГа) (USSR)—two spacecraft, each of which deployed a lander and a balloon in 1985 to study Venus before flying by Comet 1P/Halley; the Soviet Union led a multinational coalition to develop, launch, and monitor these missions; Russian contraction of Венера and Галлея, meaning Venus and Halley. L:12-15-1984 (Vega 1), 12-21-1984 (Vega 2), S:03-06-1986 (Vega 1), 03-09-1986 (Vega 2), E:01-30-1987 (Vega 1), 03-24-1987 (Vega 2) (Chapter 13).

Venera (Венера) (USSR)—sixteen missions launched between 1961 and 1983 by the Soviet Union to Venus; 4 is the first probe to land on Venus (10-18-1967); 8 also successfully landed (07-22-1972); 9 detected the acidic composition of the atmosphere (1975); 11 found sulfur and chlorine in the clouds (1978); Russian for Venus (Chapter 6).

FIGURE A-14   Venera 7 probe to Venus.

Venus Express (ESA)—orbiter to Venus; transmitted data about the upper atmosphere and its interaction with the solar wind from 2006 to 2014; showed changing nighttime cloud features driven by cooling atmosphere. L:11-09-2005, S:04-11-2006, E:12-16-2014 (Chapter 6).

Viking 1 & 2 (NASA)—a pair of orbiters and landers to study Mars; the orbiters returned extensive imagery of the planet, while the landers provided a detailed look at martian weather; the seismology experiments did not work as anticipated; the experiments designed to look for microscopic life were inconclusive. L:08-20-1975 (V1), 09-09-1975 (V2), S:07-20-1976 (V1), 08-07-1976 (V2), E:11-11-1982 (V1), 04-12-1980 (V2) (Chapter 8).

Voyager 1 (NASA)—space probe that executed flybys of the Jupiter and Saturn systems and explored the heliosphere (mini-Grand Tour); transitioned to an interstellar mission, currently outside the heliosphere and on path to leave the Solar System—the most distant human-made craft from Earth. L:09-05-1977, S:03-05-1979 (Jupiter), S:11-12-1980 (Saturn) (Chapter 10).

Voyager 2 (NASA)—space probe that completed the outer-planet Grand Tour: the Jupiter, Saturn, Uranus, and Neptune systems; currently outside the heliosphere and on an interstellar mission, leaving the Solar System. L:08-20-1977, S:07-09-1979 (Jupiter), S:08-26-1981 (Saturn), S:01-24-1986 (Uranus), S:08-25-1989 (Neptune) (Chapters 10, 11).

FIGURE A-15   Voyager 2 spacecraft, bound for the Grand Tour of the outer Planetary System.

Wilkinson Microware Anisotropy Probe [WMAP] (NASA)—spacecraft that mapped the cosmic background radiation from the Big Bang to determine the expansion rate and age of the Universe; follow-up to the COBE mission; named in honor of cosmologist David Wilkinson. L:06-30-2001, S:07-02-2001, E:08-19-2010 (Chapter 1).

Appendix V Biographies

Adams, John ⟨**1819–1892**⟩ English astronomer who predicted the position of Neptune based on its apparent gravitational influence on Uranus; Urbain Le Verrier made a similar prediction independently

Allen, Paul ⟨**1953–2018**⟩ American entrepreneur and philanthropist, best known for co-founding Microsoft Corporation with Bill Gates; funded the first private manned space plane, SpaceShipOne; founder of Stratolaunch Systems to promote space travel through cost reduction and on-demand access

Ansari, Anousheh ⟨**1966–**⟩ Iranian-American electrical engineer and the first female space tourist; founder of the Ansari XPRIZE for accomplishments in space entrepreneurism

Aristarchus ⟨*circa* **230 BCE**⟩ Greek astronomer who estimated the relative sizes of the Sun, Earth, and Moon; concluded that the Sun is bigger than the Earth and posited a heliocentric system nearly 2 millennia before Copernicus's mathematical treatment thereof

Aristotle ⟨**384–322 BCE**⟩ Greek philosopher who established the traditions of Western science; among these was a theory of uniform circular motion for celestial objects

Barnard, Edward Emmerson ⟨**1857–1923**⟩ American astronomer who, in 1916, reported that the star that now bears his name was moving extremely fast with respect to the background stars; successful comet hunter; created detailed photographic atlas of the Milky Way Galaxy

Basri, Gibor ⟨**1951–**⟩ African-American astronomer best known for his work helping to prove the existence of brown dwarfs; participated in the highly successful Kepler Mission to discover exoplanets

Bayer, Johann ⟨**1572–1625**⟩ German attorney and astronomer whose star atlas *Uranometria* ⟨1603⟩ introduced a nomenclature system for identifying stars; that system still is in use

Beebe, Reta ⟨**1936–**⟩ American planetary astronomer who describes long-term atmospheric phenomena seen in images of the giant planets made at Earth-based telescopes; interpreted features imaged by the Voyager missions as part of its Imaging Team

Bezos, Jeff ⟨**1964–**⟩ American computer scientist who founded the Amazon corporation; his Blue Origins company seeks to substantially lower the cost of space travel

Borisov, Gennadiy ⟨**1962–**⟩ Ukrainian optical engineer and astronomer who discovered the first interstellar object definitively a comet in nature, 2I/Borisov ⟨2019⟩

Brahe, Tycho ⟨**1546–1601**⟩ Danish astronomer who compiled highly accurate observations of the positions of celestial bodies; his observations of a comet in 1577 proved that it was farther away than the Moon, which contradicted the teachings of Aristotle

Branson, Richard ⟨**1950–**⟩ English businessman active in developing commercial space travel; first to cross the Atlantic and Pacific Oceans by balloon

Brown, Michael ⟨**1965–**⟩ American astronomer best known for his discovery of Eris, a massive KBO, and Sedna, a possible member of the inner Oort Cloud; knowledge of these and similar objects motivated the IAU to define planet and dwarf planet formally

Bruno, Giordano ⟨**1548–1600**⟩ Italian priest convicted of heresy and burned at the stake; claimed that stars are distant suns

Buddha ⟨*circa* **5th century BCE**⟩ née Siddhartha Gautama; Indian philosopher who demonstrated how to escape the suffering inherent in endless cycles of rebirth; founder of a major world religion; Buddha is a title meaning 'Awakened One' in Sanskrit

Carruthers, George ⟨**1939–2020**⟩ African-American physicist and inventor responsible for developing ultraviolet cameras and spectrometers for astronomy, including those used by the NASA Apollo missions

Cassini, Giovanni ⟨1625–1712⟩ Italian-French astronomer who reported the bands of Jupiter, measured rotation periods for Mars and Jupiter, found four moons of Saturn, and identified the Cassini Division in the rings of Saturn

Cecchi, Filippo ⟨1822–1887⟩ Italian physicist and instrument maker who is credited with developing the first modern seismometer to record earthquakes; a priest and member of the Clerks Regular of the Pious Schools

Colombo, Giuseppe 'Bepi' ⟨1920–1984⟩ Italian mathematician who explained Mercury's pattern of three rotations about its axis to two revolutions about the Sun as a resonance; proposed that NASA use Venus's gravity to assist Mariner 10 in achieving an orbit that would allow it to encounter Mercury multiple times

Copernicus, Nicolaus ⟨1473–1543⟩ Polish canon and astronomer who developed a heliocentric model of the Solar System and published it as *De revolutionibus orbium coelestium* ⟨1543⟩; although his was no more accurate than the Ptolemaic Model, which it eventually replaced, it fundamentally changed the way we think about the Earth and its place in the Universe

Crutzen, Paul ⟨1933–⟩ Dutch chemist who shared the 1995 Nobel Prize in Physics with Mario Molina and F. Sherwood Rowland for their research in atmospheric chemistry, especially ozone

Davis, Raymond, Jr. ⟨1914–2006⟩ American chemist and physicist who built the first device to detect neutrinos from the Sun; the result was experimental proof that the Sun shines through the power of nuclear fusion, work which earned him the 2002 Nobel Prize in Physics

Doppler, Christian ⟨1803–1853⟩ Austrian physicist who noticed his eponymous effect in sound waves; now often referred to as the Doppler-Fizeau Effect, it was ultimately applied to light and used to determine the radial velocities of astronomical objects

Drake, Frank ⟨1930–⟩ American radio astronomer and SETI pioneer; established framework for quantifying number of possible extraterrestrial civilizatons; led first scientific effort to detect alien radio signals

Eddington, Arthur ⟨1882–1944⟩ English physicist who calculated how energy is transported through stars and coordinated 1919 eclipse expeditions to test the gravitational deflection of light predicted by Einstein's General Theory of Relativity; a noted pacifist during a time of beating war drums

Edgeworth, Kenneth ⟨1880–1972⟩ Irish soldier and economist who hypothesized the existence of a host of comet-producing objects beyond the orbit of Neptune; wrote on the origins and evolution of the Planetary System

Einstein, Albert ⟨1879–1955⟩ German-American physicist who formulated our modern understanding of mass, dynamics, and gravitation; received the 1921 Nobel Prize in Physics

Elliott, James ⟨1943–2011⟩ American astronomer who used stellar occultations to discover the rings of Uranus and establish the presence of an atmosphere on Pluto; teacher of Thomas Hockey

Epicurus of Samos ⟨*circa* 341 BCE–*circa* 271 BCE⟩ Greek philosopher who formulated the view that the Universe is infinite and eternal; therefore, extraterrestrial life and intelligence must also exist

Eudoxus ⟨*circa* 390 BCE–*circa* 338 BCE⟩ Greek mathematician who proposed the Celestial Spheres Model of the Solar System; while not very predictive, it was rationally based and quantifiable

Flamsteed, John ⟨1646–1719⟩ first English Astronomer Royal, who prepared the stellar catalog *Historia Coelestis Britannica* ⟨1725⟩ using a telescope; it is the source of the so-called Flamsteed numbers for stars, e.g., 51 Pegasi is the 51st star listed in the constellation Pegasus

Galilei, Galileo ⟨1564–1642⟩ Italian physicist and astronomer who championed the Copernican Model of the Solar System; using a telescope, he saw relief on the Moon, spots on the Sun, the phases of Venus, the satellites of Jupiter that bear his name, and the stellar nature of the Milky Way; his observations contradicted the theories of his time

Galle, Johann ⟨1812–1910⟩ German astronomer who first viewed Neptune through a telescope close to the position predicted by Urbain Le Verrier; determined the astronomical unit using asteroid observations, the most accurate value at the time

Gan De ⟨*circa* **365 BCE**⟩ Chinese astronomer who developed the concept of synodic period; his recorded observations of Jupiter may include a sighting of Ganymede

Hall III, Asaph ⟨**1829–1907**⟩ American astronomer who discovered the moons of Mars; proposed modification of Newton's Laws to account for anomalies in the orbit of Mercury, which were later explained by Albert Einstein

Halley, Edmond ⟨**1656–1742**⟩ English Astronomer Royal who urged international effort to determine the Earth-Sun distance based on transits of Venus; predicted the return in 1759 of the periodic comet that bears his name

Hawking, Stephen ⟨**1942–2018**⟩ English physicist who, among other achievements in the study of gravity, elucidated the Big Bang; with funding from Yuri Milner, his Breakthrough Listen is a SETI effort while his Breakthrough Starshot seeks to send a tiny spacecraft to Alpha Centauri

Herschel, Caroline ⟨**1750–1848**⟩ German-English musician and astronomer who assisted her brother, William Herschel, at the telescope; personally discovered eight comets plus many star clusters and nebulae; published an updated and expanded star catalog

Herschel, (Frederick) William ⟨**1738–1822**⟩ German-English musician and astronomer who was assisted in his investigations by his sister, Caroline Herschel; discovered Uranus and preferred 'asteroid' for small, rocky bodies between the orbits of Mars and Jupiter; attempted to map the Milky Way Galaxy

Hertzsprung, Ejnar ⟨**1873–1967**⟩ Danish astronomer who recognized two groups of stable stars when plotting stellar temperature versus luminosity; the discovery was contemporaneous with similar work of Henry Russell, and the resulting graph is now called the Hertzsprung-Russell Diagram

Hodges, Ann ⟨**1920–1972**⟩ American, only person known to have been struck by a meteorite; a piece of the Sylacauga meteorite crashed through the ceiling of her home, hit a radio, and bruised her

Horrocks, Jeremiah ⟨*circa* **1618–1641**⟩ English astronomer who first predicted and observed a transit of Venus (4 December 1639); admired by Isaac Newton

Hubble, Edwin ⟨**1889–1953**⟩ American astronomer who classified the galaxies into spirals, ellipticals, and irregulars; correlated galactic redshifts (velocities) with distances to demonstrate that the Universe appears to be expanding

Huygens, Christiaan ⟨**1629–1695**⟩ Dutch astronomer who identified the rings of Saturn as such and discovered the Saturnian moon, Titan; recorded the first surface feature on Mars and used it to show that the planet rotated

Itokawa, Hideo ⟨**1912–1999**⟩ Japanese aerospace engineer who initiated his country's space program; his rockets reached space in 1960; launched the first Japanese satellite in 1970

Jackson, William ⟨**1936–**⟩ African-American astrochemist, specializing in the chemistry of comets, particularly laboratory work on the breakup of molecules when exposed to sunlight

Jewitt, David ⟨**1958–**⟩ English-American astronomer who, with Jane Luu, located the first KBO besides Pluto (and Charon); also spotted the second and many more

Kajita, Takaaki ⟨**1959–**⟩ Japanese physicist who determined that neutrinos come in three different 'flavors' and have mass; his work was awarded the 2015 Nobel Prize in Physics

Kepler, Johannes ⟨**1571–1630**⟩ German astronomer and early Copernican who established that the planets orbit the Sun in ellipses with the Sun at one focus, move at uneven speeds, and exhibit a relationship between their period of revolution and average distance from the Sun; began his career in earnest as assistant to Tycho Brahe

King, David ⟨**1962–**⟩ American mechanical engineer and former NASA center director now involved in commercial space endeavors with the Dynetics Corporation

Kirkwood, Daniel ⟨**1814–1895**⟩ American astronomer who determined that the Asteroid Belt contains gaps where erstwhile asteroids would be in unstable resonance with Jupiter; these radii from the Sun are known as Kirkwood Gaps

Knapp, Michelle ⟨1974–⟩ American owner of the red 1980 Chevy Malibu when the Peekskill meteorite struck it

Kopff, August ⟨1882–1960⟩ German astronomer who discovered a number of comets and asteroids, including Patroclus the second Trojan of Jupiter

Koshiba, Masatoshi ⟨1926–2020⟩ Japanese physicist who, along with Raymond Davis, Jr., detected neutrinos emitted by the Sun; awarded the 2002 Nobel Prize in Physics

Kuiper, Gerard ⟨1905–1973⟩ Dutch-American planetary scientist who determined the atmospheric composition of Mars and Titan and discovered satellites of Uranus and Neptune; believed that *few* solar-system bodies exist beyond the orbit of Neptune—nevertheless, the great reservoir of such objects inexplicably is called the Kuiper Belt

Lagrange, Joseph-Louis ⟨1736–1813⟩ Italian-French mathematician who computed that, when one body orbits another, five points exist at which the gravitational attraction of the two is equal; the Trojan asteroids are associated with the two most stable of these points; contributed to the earliest expression of the metric system (now Système International) during the French Revolution

Leavitt, Henrietta ⟨1868–1921⟩ American astronomer who established the relationship between luminosity and pulsation period of Cepheid variable stars allowing their use as standard candles to measure distances to other galaxies; her life as a deaf, female astronomer is dramatized in the 2015 play, *Silent Sky*, by Lauren Gunderson

Lemaître, Georges ⟨1894–1966⟩ Belgian priest and cosmologist known as the father of the Big Bang theory; theoretical work describes the expanding Universe; studied behavior of magnetic particles in terrestrial and galactic magnetic fields

Le Verrier, Urbain ⟨1811–1877⟩ French specialist in orbits, or the study of 'celestial mechanics'; postulated (incorrectly) a planet interior to Mercury in order to explain its anomalous precession and, independently of John Adams, predicted (correctly) the position of yet-unseen Neptune

Liu Hsin ⟨*circa* 50–*circa* 23 BCE⟩ Chinese scholar who popularized the use of year names based on Jupiter's motion through 12 Chinese zodiac constellations, as well as five 'elements'; proposed a rule adjusting the resulting 60-year cycle every 144 years because Jupiter's orbital period is less than 12 years

Lockyer, J. Norman ⟨1836–1920⟩ English astronomer who identified helium in the solar spectrum before it had been identified on Earth; traveled the world observing at least 10 solar eclipses; also discovered how to observe solar prominences when the Sun was not eclipsed

Lomonosov, Mikhail ⟨1711–1765⟩ Russian scientist and poet who detected the atmosphere of Venus during its 1761 transit; theorized on the structure of comets

Lowell, Percival ⟨1855–1916⟩ American astronomer who initiated the search for what is now Pluto at his Lowell Observatory; thought he observed Martian features that he interpreted as evidence for an alien civilization

Luu, Jane ⟨1963–⟩ Vietnamese-American astronomer who, with David Jewett, located the first KBO besides Pluto (and Charon); searched for 6 years before succeeding

Mayor, Michel ⟨1942–⟩ Swiss astronomer who, with Didier Queloz, detected the first exoplanet orbiting a Sun-like star, 51 Pegasi; they shared the 2019 Nobel Prize in Physics for this discovery

McDonald, Arthur ⟨1943–⟩ Canadian physicist showed that there are three kinds of neutrinos, all of which have mass; he shared the 2015 Nobel Prize in Physics for this work, along with Takaaki Kajita

Milankovitch, Milutin ⟨1879–1958⟩ Serbian physicist who found relationships between long-term climate change on the Earth and astronomical patterns of varying eccentricity, obliquity, and precession; these are now known as Milankovitch Cycles; proposed reform of Gregorian calendar partially accepted by some Orthodox Christian sects

Milner, Yuri ⟨1961–⟩ Russian-Israeli physicist and investor; his Breakthrough Starshot is a program to establish the viability of the light sail for solar-system propulsion; he also sponsors Breakthrough Listen, a SETI initiative

Molina, Mario ⟨**1943–**⟩ Mexican-American chemist who shared the 1995 Nobel Prize in Physics with Paul Crutzen and F. Sherwood Rowland for their research in atmospheric chemistry, especially regarding ozone

Musk, Elon ⟨**1971–**⟩ South African-American economist and physicist whose SpaceX currently is responsible for transporting astronauts into space; builds the Tesla automobile

Oort, Jan ⟨**1900–1992**⟩ Dutch astronomer who, based on the aphelion of many long-period comets, envisioned a Planetary System surrounded by a shell of potential comet nuclei at great distance from the Sun

Payne-Gaposchkin, Cecilia ⟨**1900–1979**⟩ English-American astronomer who found that stars are mainly composed of hydrogen and helium; developed a system for determining a star's temperature

Penzias, Arno ⟨**1933–**⟩ German-American radio astronomer who, with Robert Wilson, first detected the CBR in 1963, at the time called the afterglow of the Big Bang; they shared the 1978 Nobel Prize in Physics

Piazzi, Giuseppe ⟨**1746–1826**⟩ Swiss-Italian priest and astronomer; discovered the first known asteroid, Ceres; demonstrated that most stars have some transverse velocity

Poincaré, Jules-Henri ⟨**1854–1912**⟩ French mathematician who inaugurated the modern study of the n-body problem, where more than two bodies gravitationally interact; showed that resulting orbits can be chaotic

Ptolemaeus, Claudius 'Ptolemy' ⟨*circa* **85–***circa* **165**⟩ Greco-Egyptian astronomer who developed a sophisticated geocentric model of the Solar System (the Ptolemaic Model) that dominated astronomy until the 17th century; he set forth his theories in a book originally titled *The Mathematical Compilation* in Greek but which is now better known as the *Almagest*

Pythagoras ⟨*circa* **570 BCE–***circa* **480 BCE**⟩ Greek mathematician and mystic known for his theorem about relating the lengths of the sides of triangles; identified Morning and Evening Stars with a single celestial body, Venus

Queloz, Didier ⟨**1966–**⟩ Swiss astronomer who, with Michel Mayor, detected the first exoplanet orbiting a Sun-like star, 51 Pegasi; they shared the 2019 Nobel Prize in Physics for this discovery

Reines, Frederick ⟨**1918–1998**⟩ American physicist who confirmed the existence of the neutrino; shared the 1995 Nobel Prize in Physics

Roche, Édouard ⟨**1820–1883**⟩ French mathematician who calculated the minimum distance at which a satellite of equal density may orbit a planet without being torn apart by tidal forces; his investigation of comets presaged the solar wind

Roman, Nancy Grace ⟨**1925–2018**⟩ American astronomer specializing in stellar properties who was a strong proponent for the Hubble Space Telescope as one of the first female executives at NASA; a later-generation orbiting observatory was rechristened the Nancy Grace Roman Space Telescope

Rowland, F. Sherwood ⟨**1927–2012**⟩ American chemist who shared the 1995 Nobel Prize in Physics with Paul Crutzen and Mario Molina for their research in atmospheric chemistry, especially regarding ozone

Russell, Henry ⟨**1877–1957**⟩ American astronomer who recognized two groups of stable stars when plotting stellar temperature versus luminosity; the discovery was contemporaneous with similar work of Ejnar Hertzsprung, and the resulting graph is now called the Hertzsprung-Russell Diagram

Sagan, Carl ⟨**1934–1996**⟩ American planetary scientist, astrobiologist, and science communicator; attributed the surface temperature of Venus to the runaway greenhouse effect and described massive dust storms on Mars; author of science fiction novel *Contact*

Sharp, Robert ⟨**1911–2004**⟩ American geologist who interpreted early spacecraft images of Mars as showing features due to fluid flow

Schröter, Johann ⟨1745–1816⟩ German magistrate and astronomer who attempted to measure Venus's rotation rate and established the presence of its atmosphere; described solar granulation

Schwabe, S. Heinrich ⟨1789–1875⟩ German pharmacist and astronomer who noticed periodicity in the number of sunspots visible at a given time; searched futilely for an intra-Mercurial planet

Shoemaker, Carolyn ⟨1929–⟩ American astronomer known for discovering at least 32 comets; with her husband, Gene Shoemaker, and David Levy, she spotted Comet Shoemaker-Levy 9, which crashed into Jupiter in 1994

Shoemaker, Eugene 'Gene' ⟨1923–1997⟩ American geologist who determined that an asteroid impact created Barringer Crater; along with his wife, Caroline Shoemaker, and David Levy, he discovered Comet Shoemaker-Levy 9, which crashed into Jupiter in 1994; established the astro-geology branch of the US Geological Survey

Simonyi, Charles ⟨1948–⟩ Hungarian-American mathematician and two-time space tourist; he was one of the early pioneers of the Microsoft Corporation

Slipher, Vesto ⟨1875–1969⟩ American spectroscopist who measured velocities of galaxies by measuring absorption lines in their spectra; similarly measured rotation of Uranus and studied rotation of Saturnian rings

Sturluson, Snorri ⟨1179–1241⟩ Icelandic poet, chieftain, and lawspeaker; his *Prose Edda* is a both a style manual for poets and a compilation of Norse mythology

Swift, Lewis ⟨1820–1913⟩ American shopkeeper and astronomer who discovered 1,248 nebulae, most of which are now known to be galaxies, and 13 comets, including Comet Swift-Tuttle in 1862, the source of the annual Perseid meteor shower

Tempel, Ernst Wilhelm ⟨1821–1889⟩ German lithographer and astronomer who discovered 5 asteroids and 21 comets, including Comet Tempel-Tuttle in 1865, which is the source of the annual Leonid meteor shower

Tito, Dennis ⟨1940–⟩ American astronautical engineer and financier; helped send US space probes, Mariner 4 and 9, to Mars; applied mathematical models to financial markets; first ticketed spaceflight passenger

Tombaugh, Clyde ⟨1906–1997⟩ American astronomer who used photography to discover the first dwarf planet, Pluto; alternately looking between two photographic plates taken at different times revealed an object that moved with respect to the background stars

Truth, Sojourner ⟨*circa* 1797–1883⟩ née Isabella Baumfree; former enslaved woman and African-American civil rights activist; the Mars Pathfinder rover is named in her honor

Tuttle, Horace ⟨1839–1893⟩ American astronomer who discovered 2 asteroids and 10 comets, including Comet Swift-Tuttle in 1862 and Comet Tempel-Tuttle in 1866, which are respectively the sources of the annual Perseid and Leonid meteor showers

Van Allen, James ⟨1914–2006⟩ American physicist who detected two regions of high-energy, charged particles trapped by the Earth's magnetic field in 1958 using instruments he designed for the first two successful US satellites, Explorer 1 and Explorer 3

van de Kamp, Piet ⟨1901–1995⟩ Dutch-American astronomer and musician best known for his early claims of detecting an exoplanet orbiting Barnard's Star; more successful in detecting unseen stellar companions to nearby stars

von Oppolzer, Theodor ⟨1841–1886⟩ Austrian astronomer who discovered asteroid Hilda (named after his daughter), which is the prototype of a group in a 3:2 resonance with Jupiter; measured longitudes of European cities in order to characterize the topography and shape of the Earth

Whipple, Fred ⟨1906–2004⟩ American astronomer who constructed the Dirty Snowball Model for comet nuclei; established relationship between many meteors and comets

Wilkinson, David ⟨1935–2002⟩ American cosmologist who was a leading scientist on COBE and WMAP, two NASA missions to map the CBR

Wilson, Robert ⟨1936–⟩ American radio astronomer who, with Arno Penzias, first detected the CBR in 1963; they shared the 1978 Nobel Prize in Physics

Wolf, Maximillian ⟨1863–1932⟩ German astronomer who pioneered photography to find 228 new asteroids, including Achilles, the first Trojan asteroid; discovered 5,000 nebulae and galaxies

Yarkovsky, Ivan ⟨1844–1902⟩ Russian civil engineer who recognized that sunlight could apply a non-Newtonian force on small asteroids, thereby making their rotations, obliquities, and orbits more complicated to compute

Zhang Heng ⟨78–139⟩ Chinese scholar who developed the first seismometer; when the instrument detected an earthquake, one of eight dragons would drop a ball into the mouth of a waiting frog indicating the direction of the triggering event and alerting the watchmen; promoted a model of the Universe in which the Celestial Sphere encircles the Earth

Zuckerberg, Mark ⟨1984–⟩ American computer scientist who founded the Facebook social media web site; he collaborates with Yuri Milner on Breakthrough Starshot, an effort to develop light-sail propulsion technology for travel through the Solar System, and Breakthrough Listen, a SETI initiative

Zupi, Giovani ⟨1589–1667⟩ Italian priest and mathematician who resolved Jupiter's belts and zones; was the first to realize that Mercury exhibits phases

Appendix VI Acronyms and Symbols

^{1}H	Proton, also hydrogen
^{2}D	Deuterium, 'heavy' hydrogen
^{3}He	'Light' helium
^{4}He	Helium
A	Area
a	Semi-major axis
APF	Automated Planet Finder
ASI	Agenzia Spaziale Italiana (Italian Space Agency)
au	Astronomical unit
c	Speed of light
CBR	Cosmic background radiation
CCD	Charged-couple device
CEO	Chief executive officer
CFC	Chlorofluorocarbon
CME	Coronal mass ejection
CNES	Centre National d'Études Spatiales (French National Center for Space Studies)
CNSA	Chinese National Space Administration
CSA	Canadian Space Agency
COBE	COsmic Background Explorer
CoRoT	Convection, Rotation, and planetary Transits
DLR	Deutsches Zentrum für Luft- und Raumfahrt (German Aerospace Center)
DNA	Deoxyribonucleic acid
E	Energy
e^+	Positron
e^-	Electron
ESA	European Space Agency
ETC	Encke-type comet
FM	Formulation [of new] mission
FTD	Florists Transworld Delivery
HARPS	High accuracy radial velocity planet searcher
HATNet	Hungarian-made Automated Telescope Network
HTC	Halley-type comet
IAU	International Astronomical Union
IM	Implementation (mission)
IPCC	Intergovernmental Panel on Climate Change
ISRO	Indian Space Research Organisation
ISS	International Space Station
JAXA	Japanese Aerospace Exploration Agency
JFC	Jupiter-family comet
K-Pg	Cretaceous-Paleogene
KBO	Kuiper Belt Object
KELT	Kilodegree Extremely Little Telescope

LRE	Labeled Release Experiment
LRO	Lunar Reconnaissance Orbiter
m	Mass
NAS	National Academy of Science (United States of America)
NASA	National Aeronautics and Space Administration (United States of America)
NEO	Near-Earth Object
NGTS	Next-Generation Transit Survey
P	Primary, Pressure
PM	Primary mission
p-p	Proton–proton
R	Radius (of a sphere)
r	Radius (e.g., an orbit)
RNA	Ribonucleic acid
Roscosmos	Roscosmos State Corporation for Space Activities (Russia)
S	Secondary, Shear
SL-9	Comet Shoemaker-Levy-9
SuperWASP	Wide Angle Search for Planets
T	Time (period of revolution)
TRAPPIST	Automated TRAnsiting Planets and PlanetesImals Small Telescope
UAESA	United Arab Emirates Space Agency
UFO	Unidentified Flying Object
US/USA	United States/United States of America
USSR	Union of Soviet Socialist Republics
UTC	Universal Time, Coordinated
V	Volume
YORP	Yarkovsky, O'Keefe, Radzievskii, and Paddack
γ	Photon
ν_e	Electron neutrino

Appendix VII Further Reading

FURTHER READING

Books

Introduction
De Pater, Imke & Lissauer, Jack J. *Planetary Science* (Updated 2nd Edition). Cambridge University Press (2015).
Dick, Steven J. *Discovery and Classification in Astronomy: Controversy and Consensus.* Cambridge University Press (2013).

Chapter 1
Brush, Stephen G. *Nebulous Earth: The Origin of the Solar System and the Core of the Earth from Laplace to Jeffreys.* Cambridge University Press (1996).
van den Heuvel, Edward. *The Amazing Unity of the Universe: And Its Origin in the Big Bang.* New York, NY: Springer (2016).

Chapter 2
Ferguson, Kitty. *Tycho and Kepler: The Unlikely Partnership that Forever Changed our Understanding of the Heavens.* London: Transworld Digital (2013).
Timberlake, Todd & Wallace, Paul. *Finding Our Place in the Solar System.* Cambridge University Press (2019).

Chapter 3
Linton, C. M. *From Eudoxus to Einstein: A History of Mathematical Astronomy.* Cambridge University Press (2004).
Rogers, Lucy. *It's Only Astrophysics.* New York, NY: Springer (2008)

Chapter 4
Maggie Aderin-Pocock, Maggie. *Book of the Moon: A Guide to Our Closest Neighbor.* New York, NY: Henry H. Abrams (2019).
Wilhelms, Don E. *To a Rocky Moon: A Geologist's History of Lunar Exploration.* Tucson, AZ: University of Arizona Press (1993).

Chapter 5
Rothery, David A. *The Planet Mercury: From Pale Dot to Dynamic World.* New York, NY: Springer (2014).
Solomon, Sean C., Nittler, Larry R., & Anderson, Brian J. [editors]. *Mercury: The View After MESSENGER.* Cambridge University Press (2019).

Chapter 6
Kragh, Helge & Pedersen, Kurt Möller. *The Moon that Wasn't: The Saga of Venus' Spurious Satellite.* New York, NY: Springer (2008).
Ottewell, Guy. *Venus: A Longer View.* Greenville, NC: Universal Workshop (2020).

Chapter 7
Dessler, Andrew. *Introduction to Climate Change* (2nd Edition). Cambridge University Press (2016).
Hazen, Robert M. *The Story of Earth: The First 4.5 Billion Years, from Stardust to Living Planet.* London: Penguin Books (2013).

Chapter 8
Howell, Elizabeth & Booth, Nicholas. *The Search for Life on Mars: The Greatest Scientific Detective Story of All Time.* New York, NY: Arcade Publishing (2020).
Sheehan, William & O'Meara, Stephen. *Mars.* Lanham, MD: Prometheus Books (2001).

Chapter 9
Alverez, Walter. *T. rex and the Crater of Doom*. Princeton, NJ: Princeton University Press (2015).
Peebles, Curtis. *Asteroids: A History*. Washington, DC: Smithsonian Institution (2000).

Chapter 10
Sheehan, William. *Saturn*. London: Reaktion Books (2020).
Sheehan, William & Hockey, Thomas. *Jupiter*. London: Reaktion Books (2018).

Chapter 11
Bell, James. *The Interstellar Age: The Story of the NASA Men and Women Who Flew the Forty-year Voyager Mission*. New York, NY: Dutton (2016).
Standage, Tom. *The Neptune File: A Story of Astronomical Rivalry and the Pioneers of Planet Hunting*. London: Walker Books (2000).

Chapter 12
Carroll, Michael. *Icy Worlds of the Solar System: Their Tortured Landscapes and Biological Potential*. New York, NY: Springer (2019).
Hall, James A., III. *Moons of the Solar System: From Giant Ganymede to Dainty Dactyl*. New York, NY: Springer (2015).

Chapter 13
Boice, Daniel C. & Hockey, Thomas. *Comets in the 21stCentury: A Personal Guide to Experiencing the Next Great Comet!* Williston, VT: Morgan and Claypool Publishers (2019).
Eicher, David J. *COMETS!: Visitors from Deep Space*. Cambridge University Press (2013).

Chapter 14
Brown, Mike. *How I Killed Pluto and Why It Had It Coming*. New York, NY: Random House (2012).
Cruikshank, Dale & Sheehan, W. *Discovering Pluto: Exploration at the Edge of the Solar System*. Tucson, AZ: University of Arizona Press (2018).
Tombaugh, Clyde W. & Moore, Patrick. *Out of Darkness: The Planet Pluto*. Harrisburg, PA: Stackpole Books (1980).

Chapter 15
Golub, Leon & Pasachoff, Jay M. *Nearest Star: The Surprising Science of our Sun*. Cambridge University Press (2014).
Tassoul, Jean-Louis & Tassoul, Monique. *Concise History of Solar and Stellar Physics*. Princeton, NJ: Princeton University Press (2014).

Chapter 16
Tasker, Elizabeth. *The Planet Factory: Exoplanets and the Search for a Second Earth*. London: Bloomsbury Sigma (2019).
Summers, Michael & Trefil, James. *Exoplanets: Diamond Worlds, Super Earths, Pulsar Planets, and the New Search for Life Beyond Our Solar System*. Washington, DC: Smithsonian Books (2018).

Chapter 17
Dick, Steven J. *Life on Other Worlds: The 20th-Century Extraterrestrial Life Debate*. Cambridge University Press (2001).
Wanjek, Christopher. *Spacefarers: How Humans Will Settle the Moon, Mars, and Beyond*. Cambridge, MA: Harvard University Press (2020).

Websites

Introduction
Picture of the Day: https://apod.nasa.gov/apod/
To see planets visible in your night sky: https://theskylive.com/planets https://www.cfa.harvard.edu/skyreport

Chapter 1
The California Academy of Sciences simulates one scenario for the formation of the Solar System: https://www.calacademy.org/explore-science/simulating-solar-system-formation
Nice Model simulation of the evolution of the outer Solar Planetary System: http://lucy.swri.edu/2018/04/24/Nice-Model.html

Chapter 2
This *National Geographic* endorsed video illustrates the scale of the Solar System: https://www.youtube.com/watch?v=Kj4524AAZdE
Magnetic fields of Solar System bodies: https://www.nasa.gov/feature/goddard/2017/nasa-investigates-invisible-magnetic-bubbles-in-outer-solar-system

Chapter 3
NASA animates Kepler's Laws here: https://solarsystem.nasa.gov/resources/310/orbits-and-keplers-laws/
Surprising Venus-Earth spin-orbit resonance: https://apod.nasa.gov/apod/ap200603.html

Chapter 4
Watch Chang'e 4 landing on the farside of the Moon: https://www.youtube.com/watch?v=JJi_YEubKCY
The rotating Moon as observed by Lunar Reconnaissance Orbiter (LRO): https://www.youtube.com/watch?v=sNUNB6CMnE8

Chapter 5
Topographical model of Mercury from MESSENGER data: https://www.nasa.gov/feature/first-global-topographic-model-of-mercury
The 2019 transit of Mercury as seen from Solar Dynamics Observatory (don't miss the next one in 2032!): https://www.jpl.nasa.gov/edu/news/2019/11/7/a-teachable-moment-in-the-sky-the-transit-of-mercury/

Chapter 6
Akatsuki images have been animated in false color to show the super-rotation of Venus and its changing clouds: http://www.isas.jaxa.jp/en/topics/002099.html
Venus flyover using Magellan radar data: https://www.youtube.com/watch?v=3xrMu3jq6P8

Chapter 7
Intergovernmental Panel on Climate Change (IPCC): https://www.ipcc.ch
Explore our planet with Google Earth: https://www.google.com/earth/

Chapter 8
JPL-Caltech pans and zooms around the highest resolution image of Mars ever made: https://mars.nasa.gov/news/8621/nasas-curiosity-mars-rover-snaps-its-highest-resolution-panorama-yet/
Simulated flyover of Olympus Mons: https://www.youtube.com/watch?v=OTazRNGXSC8

Chapter 9
NASA map of bolides (commonly referred to as fireballs), showing a random distribution around the globe: https://www.jpl.nasa.gov/news/news.php?feature=4380
NASA's Sentry System tracks the risks of newly found NEOs: https://cneos.jpl.nasa.gov/sentry/
Online impact simulator: (Go ahead, play with it!) http://simulator.down2earth.eu/planet.html?lang=en-US
An Earth Impact Database maintained by the Planetary and Space Science Centre, University of New Brunswick, Canada: http://passc.net/AboutUs/index.html

Chapter 10
Watch the Juno orbiter whiz over Jupiter's cloud tops: https://apod.nasa.gov/apod/ap190205.html
Rotation of the Great Red Spot as seen by Juno: https://www.youtube.com/watch?v=zK4_fWbxHsI

Chapter 11
Time-lapse of "The Ten Biggest Storms We've Ever Seen in the Solar System": https://www.youtube.com/watch?v=MmGlg-tVk0E

Ride along with Voyager 2 to the Gas and Ice Giants: https://solarsystem.nasa.gov/resources/317/voyage-of-discovery/?category=planets_uranus

Chapter 12

Juno approaches Jupiter—a sped-up view of the Galilean Satellites in orbit about the giant planet: https://solarsystem.nasa.gov/moons/jupiter-moons/in-depth/

Neptune's moons and a flight over Triton: https://gravitysimulator.org/solar-system/the-neptunian-system https://solarsystem.nasa.gov/resources/337/flight-over-triton/?category=planets_neptune

Chapter 13

Comet definition survey: https://www.surveymonkey.com/r/6X2XV57

Citizen science project for analyzing captured dust from the Stardust Mission: http://stardustathome.ssl.berkeley.edu

Comets currently visible in our night skies: https://in-the-sky.org/newsindex.php?feed=comets&year=2018&month=9&day=5&town=1690313 https://theskylive.com/comets

Chapter 14

The New Horizons encounter with Arrokoth, narrated by Stephen Hawking: https://www.nasa.gov/feature/a-prehistoric-puzzle-in-the-kuiper-belt

Follow the orbits of Kuiper Belt Objects: http://pluto.jhuapl.edu/Arrokoth/About-the-Kuiper-Belt.php

Chapter 15

The Sun rotates under the view of the Solar Dynamics Observatory: (10 years condensed to 1 minute) https://www.youtube.com/watch?v=l3QQQu7QLoM&feature=youtu.be

Real-time sunspots, solar flares, and CMEs: https://www.spaceweatherlive.com/en/solar-activity

Chapter 16

This discipline is changing so rapidly that progress is best monitored online: https://exo.mast.stsci.edu https://exoplanetarchive.ipac.caltech.edu

NASA's Eyes on Exoplanets mobile app: https://eyes.nasa.gov/eyes-on-exoplanets.html

Chapter 17

The story of the first space tourist: https://www.youtube.com/watch?v=JzeYsRt7axc

The latest Planetary Science Decadal Survey (2013-2022): https://solarsystem.nasa.gov/science-goals/about/

Board Games

Exoplanets by Przemyslaw Swierczynski, Greater Than Games, LLC (2017) Create an entire planetary system, mine and terraform its planets, and seed and evolve life forms into new species. If you like it, the supplement, Exoplanets: The Great Expanse, is available.

Stellar Horizons by Dr. Andrew Rader, Compass Games (2020) Expand across the Solar System in humanity's first steps to the stars. This game is true to the laws of orbits and can be played either co-operatively or competitively. It is designed by a real-life space engineer specializing in long-duration spaceflight.

Terraforming Mars by Jacob Fryxelius, Stronghold Games (2017) Taming the Red Planet! Can you transform Mars into a habitable world? This game is recommended by Dr. Robert Zubrin, President of the Mars Society. There are many supplements to enhance your gaming experience.

Xtronaut 2.0: The Game of Solar System Exploration by Dr. Dante Lauretta, Xtronaut Enterprizes (2020) Develop realistic space missions to explore our Solar System using rocket science, politics, and strategy that mission managers face. The designer is a professor of planetary science and the Principal Investigator for the OSIRIS-REx Mission.

Glossary

A

Abiotic—without life

Acceleration—positive or negative change in the velocity; speed up, slow down, or change direction; includes deceleration

Accretion—nongravitational accumulation, e.g., dust in the proto-solar nebula sticking together

Acidophile—organism well evolved to live in an acidic environment

Albedo—measure of how well a surface reflects light, ranging from 1 (perfect reflector) to 0 (nothing reflected); light reflected divided by light received

Alkaliphile—organism well evolved to live in an alkaline environment

Anti-tail—projection of a comet's tail onto the sky such that its tip seems to point in the direction opposite the main appendage

Amino acid—organic molecule from which proteins are assembled

Anaerobe—organism well evolved to live in an oxygen-free environment

Annular—ring shaped, e.g., an eclipse in which a circular band of photosphere remains visible even while the center of the Moon's disk passes over that of the Sun's

Anorthosite—igneous rock composed mostly of the mineral plagioclase feldspar

Anthropocene Epoch—proposed unit of geologic time in which humans have significant impact on the Earth's climate and ecosystem; most recent of such units

Antimatter—atoms made out of antiparticles

Antiparticle—subatomic particle that differs only from a counterpart with an equal and opposite electrical charge; particle-antiparticle pairs annihilate each other, producing photons

Aphelion—point in a body's orbit about the Sun at which it is furthest from the barycenter ⬛ from *Helios*, the Greek god of the Sun

Apogee—furthest distance an Earth-orbiting object reaches from its barycenter ⬛ from the Greek *apogaion* for 'away from (the) Earth'

Apollo [asteroid] – category of of Near-Earth Asteroids with perihelion distance only slightly less than 1 au

Apollo [space program]—NASA series of manned missions to the Moon ⟨1969–1972⟩

Archaea—single-cell domain of life, without a well-defined nucleus; distinct from bacteria in the construction of its cell wall

Arroyo—normally dry riverbed

Artemis Accords—2020 NASA statement of principles and processes intended to govern the USA and other countries in their exploration of the Moon

Ascending node (see Node)

Asteroid—rocky irregular planetesimal orbiting the Sun that has a diameter larger than about 1 meter (40 inches) ⬛ from the Greek *asteroeides* for 'star-like'

Asteroid Belt—region of the Solar System between the orbits of Mars and Jupiter, about 2–3 au from the Sun; home to millions of asteroids; also called Main Asteroid Belt

Astrometric Method—detection of an exoplanet by the gravitational effect it has on the transverse motion of a star

Astronomical Unit (au)—average distance between the Sun and the Earth; unit of measurement for interplanetary distances = 149,597,870.700 kilometers (92,955,807.273 miles)

Aten—category of Near-Earth Asteroids with aphelion distance only slightly less than 1 au

Aurora (aurorae)—sky glow produced when an atmosphere becomes energized by extraterrestrial, subatomic particles channeled into it by its magnetic field ✍ named for the Roman goddess of the dawn

Autumnal Equinox (see Equinox)

B

Bacteria—single-cell domain of life, without a well-defined nucleus; distinct from archaea in the construction of its cell wall

Barycenter—center of mass of a multibody orbital system; each object moves around this point at a distance that is inversely proportional to its mass

Basalt—igneous rock composed of the minerals plagioclase, pyroxene, and olivine

Basin—depressed landform produced by an asteroid-mass body striking a planetary or satellite surface

Belt—low-albedo cloud band encircling a Jovian Planet parallel to its equator

Big Bang—creation event for the Universe; Universe begins to expand

Binary asteroid—two asteroids of similar mass in orbit about a common barycenter

Binary star—two stars in orbit about a common barycenter

Black drop effect—during a transit, the distortion of Venus's silhouette as it first contacts the Sun's limb; makes timing this instant difficult

Bolide—bright, sporadic meteor, sometimes called a "fireball"

Breccia—composite rock type consisting of angular fragments and other material

Brown dwarf—intermediate-mass body between that of a planet and a star; only minor nuclear reactions take place in its core

Butterfly Effect—idea that a small perturbation in a chaotic system may result in large effects elsewhere in the system

C

Caldera—round depression caused by volcanic activity ✍ from the Latin *caldarium* for 'hot bath'

Cambrian Explosion—tremendous diversity of life that appeared during this geologic period from about 541 to 485 million years ago

Cassini Division—major gap in the rings of Saturn kept clear by orbital resonances; similar to the Kirkwood Gaps in the Asteroid Belt

Catena (catenae)—crater chain on a planetary or satellite surface

Celestial—having to do with the sky

Celestial Mechanics—mathematical specialty within astronomy that deals with orbits

Celestial Sphere—imaginary hollow sphere surrounding the Earth onto which all celestial objects are projected

Centaur—small body with characteristics of both asteroids and comets, revolving about the Sun between the orbits of Jupiter and Neptune

Centrifugal force—apparent force outward from a center of curvature experienced by a body that is not traveling transversely at a constant speed

Chaotic—moving in a manner such that the body's future position and velocity cannot be determined from its present position and velocity

Chondrite—stony meteorite that has not been subject to geologic activity (e.g., melting)

Chromosphere—middle layer of the 'solar atmosphere'; seen during a total solar eclipse, it glows with the characteristic color of hot hydrogen gas

Coma (comae)—constantly escaping cloud of gas and dust surrounding an active comet; comet's 'atmosphere' ✍ from the Greek *komē* for 'hair of the head'

Comet—small undifferentiated body of rock and ice, which orbits the Sun; may pass by the Sun only once or travel through the Planetary System periodically ✍ from the Greek *komētēs* for 'long-haired (star)'

> **Short-period comet**—comet (1) with an orbital period less than 200 years, (2) low inclination, and (3) generally prograde motion; most originate in the Kuiper Belt and Scattered Disk
> **Long-period comet**—comet with (1) an orbital period exceeding 200 years, up to about a million years, (2) a random inclination, and (3) a random direction of revolution

Complex crater (see crater)

Concentric spheres—classical Greek model of the Planetary System in which the Earth is surrounded by nested, turning, transparent spheres, upon each of which is mounted the Moon, the Sun, or one of the naked-eye planets

Contact binary—two objects orbiting a common barycenter while in physical contact with each other

Convection—heat transfer the occurs within a gravitation field as parcels of fluid rise and fall

Copernican Model—Renaissance model of the Planetary System in which the planets, including the Earth, orbit the Sun, while the Moon orbits the Earth

Core [planet or satellite]—central, differentiated mass

> **Inner core**—higher-density layer of a planetary core, within an outer core; the two are often distinguished by state
> **Outer core**—lower-density layer of a planetary core, concentric with an inner core; the two are often distinguished by state

Core [Sun]—central region in which nuclear fusion takes place

Corona (coronae)—concentric oval features on the surface of Venus, probably due to rising subsurface magma

Corona [Sun]—hot but tenuous gas of electrically charged particles surrounding the Sun and extending millions of kilometers into space; seen during a total solar eclipse ✍ from the Latin for 'crown'

Coronal Mass Ejection—release from the Sun of a large plasmoid; if it happens to intersect the Earth, a geomagnetic storm is possible

Cosmic Background Radiation—electromagnetic radiation produced in the Big Bang, now permeating the Universe in radio wavelengths

Cosmogony—study of the origin and development of the known Universe

Cosmology—study of the structure and evolution of the known Universe

Crater—round depression caused by the explosive collision of a smaller body with a larger one ✍ from the Greek for 'cup'

> **Complex crater**—impact feature larger than a simple crater, characterized by terraced walls, a central peak, or a ring of central mountains
> **Simple crater**—bowl-shaped impact feature; smaller than a complex crater

Crescent (see Phase)

Cretaceous—geologic period from 140 million to 65 million years ago during which dinosaurs continued to flourish until their extinction at the end of this period while mammals and flowering planets appeared ✍ from the Greek *crēta* for chalk

Crust—outermost, solid layer of a differentiated body

Cryophile—organism well evolved to live in an extremely cold environment

Cryovolcano—eruption of volatiles from just below a cold surface within the Planetary System; similar to a geyser ✍ from the Greek *kryos* for 'icy cold'

C-type—carbonaceous asteroid made of clay and silicates

D

Decadal Survey on Planetary Science—NAS document stating research priorities in planetary science for a 10-year period; sponsored by the US National Science Foundation, US Department of Education, and NASA

Deceleration (see Acceleration)

Deferent—geocentric circle about which the center of the epicycle revolves in the Ptolemaic Model

Deoxyribonucleic Acid [DNA]—long, complex molecule found in living cells that encodes genetic information

Deposition—action of depositing material

Descending node (see Node)

Deuterium—hydrogen isotope with both a proton and neutron in its nucleus; "heavy" hydrogen

Differentiation—materials of different densities separate into layers in the presence of gravity

Direct Imaging Method—detection of an exoplanet by resolving it from its parent star; difficult because of the glare produced by the star

Dirty Snowball Model—amorphous, undifferentiated accretion of ice and dust; theory for the structure of a comet nucleus

Disconnection event—cometary gas tail separates from the coma of a comet

Diurnal—having to do with the rotation of the Earth

Doppler Effect—change in wavelength due to the source and observer moving radially with respect to each other; affects both sound and electromagnetic waves

Drake Equation—thought experiment by which the number of intelligent extraterrestrial civilizations now occupying the Galaxy might be estimated

Dust—small grains composed of materials that retain their form and properties at high temperatures, such as rocks and minerals; some with ices

Dust devil—small atmospheric vortex

Dust tail (see Tail)

Dwarf planet—body that fails only the third definitional condition for a planet, it orbits the Sun and has sufficient mass for its gravitation to compact it into a round shape but their orbital region contains many similar bodies

Dynamics—study of motion

Dynamo (see Magnetic dynamo)

E

Eccentricity—measure of how out of round an elliptical orbit is; eccentricities are greater than 0 (a circle) but less than 1 (a parabola)

Eclipse—disk of one body in the sky blocks light from another body of similar apparent size, e.g., solar or lunar eclipses

Ecliptic—plane of the Earth's orbit about the Sun ⊶ from the Greek *ekliptikos* for 'of an eclipse'

Electric field—region in which an electric force is exerted

Electromagnetic radiation—energy transported by photons; light

Electromagnetic spectrum—entire range of electromagnetic radiation, or light, from radio waves to gamma rays

    Radio—wavelengths longer than 1 meter (39 inches); can penetrate cloud layers

    Microwave—wavelength range between 1 meter and 0.001 meters (39 and 0.039 inches); used in microwave ovens to excite water molecules

Infrared—wavelength range between 0.001 meters and 750 nanometers (0.039 and 3.0×10^{-5} inches); just slightly longer than human eyes can detect

Visible—wavelength range between 750 and 380 nanometers (3.0×10^{-5} and 1.5×10^{-5} inches); wavelengths to which human eyes are sensitive

Ultraviolet—wavelength range between 380 and 10 nanometers (1.5×10^{-5} and 3.9×10^{-7} inches); just slightly shorter than human eyes can detect

X-ray—wavelength range between 10 and 0.01 nanometers (3.9×10^{-7} and 3.9×10^{-10} inches); can penetrate flesh

Gamma ray—wavelengths less than 0.01 nanometers (3.9×10^{-10} inches); can penetrate bone

Electrostatic—having to do with stationary electric charge

Ellipse—closed curve path with two foci; for each point on the curve, the sum of the distances to each focus is constant

Elongation—angle between the Sun and a planet or planet and its moon, as observed from the Earth; measured 0–180° east or west of the primary body

Encke-type—comets with the shortest known periods, the orbits of which do not reach Jupiter

Endolith—organism well evolved to live in a high-pressure environment, e.g., inside rock

Epicycle—circle upon which a planet, the Sun, or the Moon is mounted in the Ptolemaic System; in turn, it is centered on the deferent

Equinox—time at which the Sun appears to cross the Celestial Equator; daylight and nighttime periods are of equal duration

Autumnal equinox—crossing traveling South; usually on 22/23 September in the Northern Hemisphere

Spring/Vernal equinox—crossing traveling North; usually on 20 March in the Northern Hemisphere

Era of Heavy Bombardment—period early in the history of the Solar System during which left-over planetesimals peppered planetary surfaces with great frequency; about 4 billion years ago

E-type—asteroid with a surface primarily composed of the mineral enstatite, such as the Hungaria asteroids

Evening star—Venus referred to as it appears in the sky after sunset

Exobiology—study of life originating beyond the Earth ⬃ from the Greek for 'outer' and for 'life'

Exocomet—comet of an exoplanet

Exomoon—satellite of an exoplanet

Exoplanet—planet orbiting a star other than the Sun

Extrasolar—beyond the Sun

Exosphere—highest layer of an atmosphere from which particles are likely to escape

Extraterrestrial—beyond the Earth

Extremophile—organism well evolved to live in an environment normally considered inhospitable to life

F

Farside—hemisphere of a satellite in synchronous orbit, facing away from its planet

First Quarter (see Phases)

Flare—sudden eruption of mass and electromagnetic radiation from the Sun

Fluorescence—invisible electromagnetic radiation is absorbed by molecules that then emit visible light

Focus—one of two points, the sum of the distances from which to a point on an ellipse remains constant

Force—push or pull that causes an acceleration

Full Moon (see Phases)

G

Galilean satellite—one of the four, largest, regular moons of Jupiter; discovered by Galileo when he first observed the planet through a telescope

Galaxy—large, rotating system of millions to trillions of stars, plus gas and dust, such as our Milky Way Galaxy ◅ from the Greek *gala* for 'milk'

Gamma ray (see Electromagnetic spectrum)

Gas Giant—planet made up principally from layers of gaseous, liquid, and metallic hydrogen; may possess a rocky/metal core

Gas tail (see Tail)

General Theory of Relativity—system of physics in which mass and energy distort space-time to produce gravity; expansion of Albert Einstein's original theory to include acceleration due to gravity

Geographic pole—intersection of a planet's rotational axis and the surface of the planet

Geologic activity—Solar-system body's ability to reshape its own surface

Geomagnetic storm—disturbance in the Earth's magnetosphere

Geophysics—physical study of the Earth, or similar body

Geosynchronous—satellite orbit such that the period of revolution equals the Earth's rotation period; the satellite appears stationary in the sky

Geothermal—warmed by internal heat from the Earth

Geyser—eruption of water into the air due to heating of the water table by magma

Gibbous (see Phases)

Glacier—landform of frozen water in motion

Goldilocks planet—planet within a star's Habitable Zone

Grand Tack Hypothesis—theory of solar-system cosmogony in which the Gas Giants migrate from the inner to outer Planetary System

Granulation—appearance of convective cells on the solar photosphere

Gravity—attractive force at a distance produced by mass; alternately, the distortion of space-time in the presence of mass

Greenhouse effect—planetary atmosphere is transparent to visible wavelengths of sunlight, but opaque to infrared; energy absorbed by the surface cannot be reradiated away as heat, so the atmosphere warms

H

Habitable Zone—distance range from a star at which water can exist in the liquid state

Halley-type—comets with periods between 20 and 200 years

Harvest Moon—Full Moon nearest the Autumnal Equinox; alternately, the first Full Moon after the Autumn Equinox

Heliopause—distance from the Sun at which the pressure of the solar wind balances that of the interstellar medium

Heliosphere—region surrounding the Solar System influenced primarily by the Sun's solar wind and magnetic field

Highlands—higher, older terrain on the Moon; contrasted with maria

Hot Jupiter—Gas Giant–mass planet near its host star with a short revolution period; found only among exoplanets

Hot Neptune—Ice Giant—mass planet near its host star with a short revolution period; found only among exoplanets

Hydrate—compound made of water and another molecule
Hydrocarbon—organic molecule composed of hydrogen and carbon
Hydrostatic equilibrium—outward pressure at every radius of a fluid sphere equals the inward pull of gravity at that radius; such that the body maintains a constant volume

I

Ice—solid made out of a volatile substance, e.g., water, carbon dioxide, *etc.*
Ice age—cool period during which a planet is substantially covered in ice; Earth has experienced several such intervals
Ice Giant—planet made up principally from layers of gaseous and liquid hydrogen plus inner layers of ice; may possess a rocky/metal core
Igneous—rock formed from solidified lava ⬭ from the Latin for 'of fire' or 'on fire'
Inclination—angle the plane of an orbit makes with respect to the ecliptic
Inertia—resistance of mass to change in velocity
Inferior conjunction—planet passes in its orbit between the Earth and Sun
Infrared (see Electromagnetic spectrum)
Inner cloud (see Oort Cloud)
Inner core (see Core)
Insolation—amount of solar radiation reaching a surface; more broadly, radiation from a host star reaching a surface
Interplanetary dust—small particles occupying the plane of the ecliptic
Ion—electrically charged particle caused by adding or removing one or more electrons from a neutral atom or molecule
Ion tail (see Tail)
Irregular satellite—moon in a retrograde and/or highly inclined orbit
Isotope—variant of a chemical element with a different number of neutrons in its nucleus but generally having the same chemical properties

J

Jet—rapid stream of gas from a point source; alters a comet's orbit as required by Newton's Third Law
Jet stream—stable east-west or west-east wind at a particular latitude
Jovian Planet—planet similar in size to Jupiter; composed mainly of hydrogen and helium
Jupiter family—comets with periods less than 20 years

K

Kepler's Laws—set of empirical descriptions of a planetary orbit; 1) planet follows an elliptical path with the Sun at one focus, 2) planet revolves such that the areas between two segments of its path and the Sun are always equal, and 3) the square of a planet's period of revolution is proportional to the cube of its average distance from the Sun
Kirkwood Gap—distance from the Sun within the Asteroid Belt kept clear by orbital resonances; similar to the Cassini Division within Saturn's rings
Kuiper Belt—region of the Solar System extending in the plane of the ecliptic from the orbit of Neptune out to about 1,000 au in which millions of icy and rocky bodies orbit; includes the dwarf planets Pluto and Eris

L

Labeled Release Experiment—watershed experiment aboard the Viking landers designed to search for signs of martian life; results inconclusive

Lagrange point—one of five associated with an orbit where the gravitational attraction of the two bodies is balanced; points in its path 60° before and after the orbiting body are the most stable

Lava—magma on the surface of a Solar-system body

Limb—visible extent of a sphere

Long-period comet (see Comet)

Luminosity—intrinsic amount of light emitted by a source; contrasts with brightness, which is dependent upon distance from the light source

Lunar—of or pertaining to the Moon, Latin-derived; more common than Selenian from the Greek ✍ from Luna, the Roman goddess of the Moon

M

Main Asteroid Belt (see Asteroid Belt)

Magma—liquid rock within a planet or satellite ✍ from the Greek for "thick unguent" or "ointment"

Magnetic dynamo—mechanism whereby currents within a conducting fluid or metal generate a magnetic field while undergoing rotation and convection; several planets and satellites, including the Earth and Jupiter, generate magnetic fields this way

Magnetic field—region in which a magnetic force is exerted

Magnetic pole—surface spot where a magnetic field is strongest

Magnetosphere—region around an astronomical body in which charged particles are affected by that body's magnetic field, e.g., the Earth

Main Sequence—combination of temperatures and luminosities appropriate for a stable star powered by fusion of hydrogen to helium

Mantle—layer of a terrestrial planet made of high-density rock; elsewhere, any differentiated layer below the crust

Mare (maria)—dark impact basins on the Moon; an asteroid or comet strike broke the lunar crust and allowed the upwelling of lava

Mascon—concentration of mass beneath a lunar mare, discovered by changes in the acceleration of orbiting spacecraft; the word is a portmanteau of 'mass' and 'concentration'

Massif—compact group of mountain-size planetary features

Mega-Earth—terrestrial exoplanet at least ten times the mass of the Earth

Megaregolith—fused, compacted aggregate of rock below the regolith on the Moon

Mesosphere—layer of a planetary atmosphere above the stratosphere; temperature decreases with altitude

Metallic hydrogen—fluid phase of hydrogen characterized by freely moving electrons and good electrical conductivity

Metallicity—fraction of a body such as a star, made up of elements other than hydrogen, helium, or lithium

Meteor—streak of light caused by a meteoroid ignited as it passes through an atmosphere ✍ from the Greek *ta meteopa* for 'high in the air'

Meteor shower—increase in the frequency of meteors appearing to come from a common point in the sky; caused by the Earth intercepting a swarm of meteoroids

Meteorite—rock specimen of extraterrestrial origin; all or part of a meteoroid that lands intact on a planetary or satellite surface

Meteoroid—small rocky, metallic, or icy piece of natural debris orbiting the Sun, ranging in size from that of a grain of sand to that of a boulder; larger than a micrometeoroid but smaller than an asteroid

Meteorology—study of weather

Microlensing Method—detection of an exoplanet by the gravitational focusing of light

Micrometeorite—grain-sized meteorite

Micrometeoroid—grain-sized meteoroid

Microwave (see Electromagnetic spectrum)

Mini-Neptune—exoplanet that is structurally and compositionally an Ice Giant, but has less than ten times the mass of the Earth

Minor Planet—small body that is neither a planet nor a comet, i.e., dwarf planet, asteroid, Trojan, Centaur, or Kuiper Belt Object

Moment of inertia—freely rotating body's resistance to change in angular velocity

Mons (montes)—planetary or satellite mountain

Morning star—Venus referred to as it appears in the sky before sunrise

M-type—asteroid made primarily of metal

N

Nanobacterium—about 1/10 the size of a normal bacterium

Neap tide (see Tide)

Nearside—hemisphere of a satellite in synchronous orbit facing toward its planet

Near-Earth Object—Solar-system body with an orbit that has a closest approach to the Sun of less than 1.3 au; includes asteroids and comets with periods less than 200 years

Nebula—interstellar cloud of dust and gas ✍ from the Latin for 'mist' or 'vapor'

Neutrino—neutral, nearly massless, subatomic particle that rarely reacts with normal matter and travels near the speed of light

Neutron star—compact end state of a high-mass star made entirely of neutrons; individual elements no longer exist within it

New Moon (see Phases)

Newton's Laws—1) every object's velocity remains constant unless acted upon by a force, 2) a force is proportional to mass and acceleration, and 3) for every force applied in one direction there is an equal force in the opposite direction; formulated by Isaac Newton

Nice Model—modern, computer-generated cosmogony that attempts to explain the positioning and distribution of the planets and small bodies within the Solar System

Node—intersection of an orbit with the plane of the ecliptic

    Ascending node—body crosses the ecliptic traveling northward

    Descending node—body crosses the ecliptic traveling southward

Nuclear energy—that released by nuclear fission, nuclear fusion, or radioactive decay

Nuclear fission—energy production by the splitting of a high-atomic-number nucleus into nuclei of lower atomic number; associated with human efforts to generate nuclear energy

Nuclear fusion—energy production by the joining of low-atomic-number nuclei into a higher atomic-number nucleus; associated with stellar cores

Nucleus [of a comet]—inner-most, solid part of a comet

Nucleus [of an atom]—combination of protons and neutrons about which electrons 'orbit'

O

Oblateness—ratio of an ellipsoidal body's minimum diameter to its maximum diameter; oblate planets tend to bulge out at the equator with larger equatorial diameters than polar diameters

Obliquity—angle between an imaginary line perpendicular to a planet's orbital plane and planet's rotation axis; larger such angles produce greater seasonal variation

Occultation—one celestial object blocks light from another

Oort Cloud—outer region of the Solar System containing potential long-period comet nuclei

    Inner Cloud—doughnut-shaped portion of the Oort Cloud, about 2,000 to 20,000 au from the Sun; it smoothly joins the Kuiper Belt and Outer Cloud

    Outer Cloud—spherical portion of the Oort Cloud, about 20,000 to 150,000 au from the Sun; it is at the effective limit of the Sun's gravitational influence

Orbit—closed path taken by one object around another due to gravity; such paths are usually elliptical

Organic—molecule containing one or more carbon atoms; associated with life

Outer core (see Core)

Outer Space Treaty—1967 international accord establishing laws governing the peaceful exploration of space; forbids military operations on solar-system bodies

P

Paleogene—period of geologic time extending from 66 until 23 million years ago; mammals flourished and diversified

Paleontologist—scientist who studies of animals and plants using their fossil record ✎ from the Greek *palaios* for 'old' or 'ancient'

Palimpsest—imprinted feature that has lost its third-dimensional component

Pangea—land mass that represented the last time a single continent existed on the Earth

Panspermia—idea that life can be transported naturally from one stellar system to another

Parallax—angle through which an object seems to move when viewed from two disparate places, as reckoned against some more-distant background; it is inversely proportional to distance

Penumbra [eclipse]—outer, incompletely shaded portion of a shadow

Penumbra [solar photosphere]—outer, less-dark region of a sunspot

Perigee—closest distance an Earth-orbiting object reaches to its barycenter ✎ from the Greek *peregeion* for "near to the Earth"

Perihelion—point in a body's orbit about the Sun at which it is closest to the Sun ✎ from *Helios*, the Greek god of the Sun

Period [of spin]—amount of time to complete one rotation

Period [of an orbit]—amount of time to complete one revolution

Permafrost—permanently frozen subsurface water

Phase—fraction of a solar-system body illuminated; although usually associated with the Moon, Mercury and Venus also go through similar phases

    Wax—to increase in fraction of surface illuminated or brightness

    Wane—to decrease in fraction of surface illuminated or brightness

    New—none of the visible surface of the body is illuminated; once new, a body transitions from waning to waxing

    Crescent—less than half of the visible surface of the body is illuminated; phase may be waxing or waning

    Quarter—half of a body's visible surface is illuminated; First Quarter occurs as the illuminated fraction waxes, while Last (Third) Quarter occurs as the illuminated fraction wanes

    Gibbous—more than half of the visible surface of the body is illuminated; phase may be waxing or waning

    Full—all the visible surface of the body is illuminated; once full, a body transitions from waxing to waning

Photo-chemical—reaction at the molecular level for which the energy derives from wavelengths of light

Photon—massless subatomic particle that carries electromagnetic energy

Photosphere—inner-most layer of the solar "atmosphere;" visible to the naked eye outside of total solar eclipse

Planet—body that 1) orbits the Sun on its own, 2) has sufficient mass to become nearly spherical under the influence of its own gravity, and 3) has cleared the region around its orbit of similar bodies ✍ from the Greek *planētēs* for 'wanderer'

Planita (planitae)—plain on a planetary or satellite surface

Planetary System—inner region of the Solar System, Mercury through Neptune

Planetesimal—primitive, accreted body massive enough for mutual gravitation attraction to cause it to interact with other planetesimals; the result of planetesimal mergers is a proto-planet

Plasma—gas-like, totally ionized state of matter

Plasmoid—'blob' of plasma held together by magnetic fields

Plate tectonics—movement of segments in a terrestrial planet's crust; responsible for some forms of geologic activity

Plutino—KBO in orbital resonance with Neptune

Polyextremophile—organism well evolved to live in two or more environments normally considered inhospitable to life

Positron—antimatter counterpart to the electron, antielectron

Prebiotic—before life

Precession—change in the orientation of a body's spin axis when disturbed by a force; the rotation axis of the Earth precesses due to gravitational torques from the Sun and Moon

Primary, pressure wave [P wave]—seismic disturbance propagated radially

Prominence—loop-like solar feature emerging from the photosphere into the corona

Proplyd (see Proto-planetary disk)

Proton-proton cycle [p-p cycle]—chain of nuclear reactions that produces energy in the Sun's core

Proto-planetary disk—distribution of material from which a planetary system formed; also called a proplyd

Proto-solar nebula—interstellar gas and dust cloud out of which the Sun and its Solar System formed some 4.6 billion years ago

Proto-star—stellar-mass body before it contracts to core temperatures and pressures necessary to initiate sustained nuclear fusion

Proto-sun—Sun as a proto-star

Ptolemaic Model—Roman-era envisioning of the Solar System in which the Sun, the Moon, and five naked-eye planets, mounted upon epicycles and deferents, orbit a stationary Earth; superseded by the Copernican Model

Q

Q-type—asteroid made primarily of the minerals olivine and pyroxene mixed with metals

R

Radar—technique for bouncing radio waves off an object and interpreting the reflected signal to determine surface features and motion (RAdio Detection And Ranging, or RADAR)

Radial Velocity Method—detection of an exoplanet by the host star's Doppler shift as it and the planet orbit their barycenter

Radio (see electromagnetic spectrum)

Radioactive—unstable elements emitting subatomic particles or photons; can result in transmutation of one element to another

Radioresistant—organism well evolved to live in a radioactive environment usually considered harmful to life

Random walk—direction of each step is independent of the previous one

Ray—one of multiple surface features radiating out from an impact crater; formed by fresh material from just below the surface being ejected during the collision

Regolith—unconsolidated rock on the lunar crust produced by impacts ✑ from the Greek *rhegos* for 'rug' or 'blanket'

Resolution—ability to distinguish two visual elements

Resonance—one body exerts a consistent, periodic, gravitational force on another

Retrograde [motion]—interval during which a planet, as viewed from Earth, appears to move westward ('backward') against the background of the Celestial Sphere

Retrograde [rotation or orbit]—in the opposite direction of most similar bodies in a system, e.g. clockwise rotation of Venus as opposed to counterclockwise motion of most planets

Revolve—travel in a closed path around a central point

Ring—orbiting particles filling, or nearly filling, a radius from a planet or other solar-system body

Rotate—spin on an axis

Rover—portion of a spacecraft designed move across a planet's or satellite's surface

Rubble pile model—amorphous, undifferentiated, collection of small bodies of ice and dust bound together only by gravity; theory for the structure of a comet nucleus

Rupes (rupes)—scarp on a planetary or satellite surface

S

Scarp—cliff produced by radial splitting of a planet's surface ✑ from 'escarpment'

Scattered Disk—thinly populated radii from the Sun overlapping the Kuiper Belt, containing potential short-period comets and Centaurs with very elliptical orbits and higher inclinations

Secondary crater—one produced by ejecta from an impact

Sedimentary rock—formed from chemistry or precipitation in a solution; deposited in one or more layers

Seismology—study of vibrations through a body, such as a planet, and their associated effects

Secondary, Shear wave [S wave]—seismic disturbance propagated transversely

Shield volcano—one formed by lava flows, named for its profile resembling a warrior's shield

Shepherd satellite—moon embedded in a ring system, the gravity of which maintains the integrity of the rings

Short-period comet (see Comet)

Sidereal—reckoned with respect to fixed points on the Celestial Sphere

Simple crater (see Crater)

Snow line—distance from a star at which water freezes

Snowball Earth—hypothesis that during times of extreme climate, the Earth was once completely covered by glaciers

Sodium tail (see Tail)

Sol—one day on Mars measured, e.g., from sunrise to sunrise; used by spacecraft-mission controllers and science fiction authors

Solar—of or pertaining to the Sun ✑ named for the Roman god of the Sun

Solar Nebula Model—theory for the origin of the Solar System first proposed in the 19th century; later backed up by computer simulations

Solar Neutrino Problem—products of nuclear fusion, the number of neutrinos detected from the Sun did not seem to match that predicted from theory; resolved by the discovery that neutrinos decay into different forms

Solar flare—burst of radiation from the Sun, localized to a sunspot group; it results in a sudden brightening on the photosphere

Solar wind—charged particles flowing outward from the Sun

Solstice—on Earth, one of two days during the year at which the Sun appears farthest from the Celestial Equator, corresponding to the days on which the Sun appears above the horizon for the longest and shortest time

> **Summer Solstice**—in the Northern Hemisphere, the day on which the Sun rises and sets farthest north and appears at its greatest noontime altitude; in the Southern Hemisphere, the day on which the Sun rises and sets farthest south and appears at its greatest noontime altitude; usually 20/21 June in the Northern Hemisphere
> **Winter Solstice**—in the Northern Hemisphere, the day on which the Sun rises and sets farthest south and appears at its least noontime altitude; in the Southern Hemisphere, the day on which the Sun rises and sets farthest north and appears at its least noontime altitude, usually 21/22 December in the Northern Hemisphere

Spicule—jet of hot gas emanating from the solar chromosphere into the corona

Spoke—low-albedo, radial feature in a set of planetary rings; caused by magnetically suspended dust

Spring tide (see Tide)

Sputtering—ejection of surface materials cause by the impact of ions

Stalactite—minerals precipitated out of a solution are deposited in a hanging structure

Standstill—farthest north or south the Moon rises during the month; similarly, the farthest north or south the Moon sets during the month

Stellar system—set of one or more exoplanets orbiting a single star or—more rarely—multiple stars

Stratosphere—layer above the troposphere in a terrestrial planetary atmosphere; temperature increases with altitude

Strong nuclear force—causation at a distance significant only on a subatomic scale; responsible for binding protons and neutrons together

S-type—asteroid made primarily of rock

Subgiant—large star above the Main Sequence on the Hertzsprung-Russell Diagram

Sublimation—phase change from solid to gas, and vice-versa

Sun-grazer—comet that comes within hundreds of thousands of kilometers of the Sun at perihelion and may plunge into it

Sunspot—region of high magnetism and reduced temperature on the Sun's photosphere

Sunspot cycle—periodic rise and fall in the number of sunspots visible on the Sun; roughly 11 years

Supergranule—meta-scale convection cell on the solar photosphere

Super-Jupiter—exoplanet more massive than Jupiter

Super-Earth—terrestrial exoplanet more massive than the Earth and of similar composition

Supergiant—among the largest and most massive stars

Supernova (supernovae)—exploding high-mass star caused by stellar collapse and rebound

Super-rotation—planetary atmosphere spins faster than the planet itself

Synchronous rotation—spin exhibited by a body, such as a planetary satellite, with a rotation period equal to its revolution period

Synodic—reckoned with respect to the Sun

T

Tail—long appendage that streams away from a comet's coma once in the inner Solar System; comets may exhibit one or more tails simultaneously
    Dust—made up of particles; affected by sunlight
    Gas/Ion—made up of atomic and molecular ions; affected by the solar wind
    Sodium—made up of neutral sodium atoms; affected by sunlight
Terminator—division between light and shadow ✍ from the Latin *terminus* for 'end' or 'limit'
Terraform—change a planetary surface environment into one resembling that of the Earth
Terrestrial—of or pertaining to the Earth; also, of or referring to the Earth-like worlds of the inner Solar System ✍ from *Terra*, the Roman goddess of the Earth
Tessera (tesserae)—small, broken piece; complex ridged landform on Venus, probably due to the crust being compressed
Third Quarter (see Phases)
Thermophile—organism well evolved to live in an extremely hot environment
Thermosphere—layer of a planetary atmosphere above the mesosphere; temperature increases with altitude
Tidal bulge—asphericity of a body due to a tidal force
Tide—distortion of a body in response to its different areas experiencing different levels of gravitational attraction to another body, high and low tides alternate four times per day on the Earth
    Spring tide—on the Earth, a higher than usual high tide, or lower than usual low tide, caused by the Moon and Sun aligning at New or Full Moon; occurs twice per month
    Neap tide—on the Earth, a lower than usual high tide, or higher than usual low tide, caused by the gravitational attraction of Moon and Sun partially canceling one another at First or Last Quarter Moon; occurs twice per month
Totality—interval during a total solar eclipse when the photosphere is completely covered by the Moon, lasting up to 7½ minutes
Trajectory—path followed by a projectile
Transit—apparent disk of a body crosses the apparent disk of another
Transit Method—detection of an exoplanet by the diminution of starlight as the planet passes between the star and Earth
Trojan asteroid—small body trapped near a Lagrange point in the orbit of a planet, about 60° ahead or behind; largest known populations are those leading (Greek node) or trailing (Trojan node) Jupiter; classified as an asteroid but may be a comet nucleus that was captured by Jupiter in the early history of the Solar System
Troposphere—layer of a terrestrial planetary atmosphere in contact with the surface; temperature decreases with altitude
Tsunami—one or more ocean waves produced by a tectonic event ✍ from the Japanese for 'harbor' and for 'waves'
Tunguska Event—1908 asteroid or comet detonation above Siberia; trees were toppled and hundreds of reindeer were killed

U

Ultraviolet (see Electromagnetic spectrum)
Umbra [eclipse]—inner, fully shaded portion of a shadow
Umbra [solar photosphere]—inner, darker region of a sunspot
Uniform circular motion—classical Greek idea that 'perfect' motion, such as that which must exist in the heavens, is required to involve only circular paths and constant angular speeds

V

Vallis (valles)—planetary or satellite valley

Van Allen Radiation Belts—two regions of high-energy, charged particles trapped by the Earth's magnetic field

Velocity—speed and direction of a moving object; an object at rest has zero velocity, that is both its speed and direction are zero

Vernal Equinox (see Equinox)

Visible (see Electromagnetic spectrum)

Volatile—under normal pressures, chemical substances that turn to gas unless kept very cold, e.g., carbon dioxide, methane, molecular oxygen, molecular nitrogen, *etc.*; water ice also is considered to be a volatile

Volcano—opening in the crust of a planet through which lava and other materials may flow onto the surface

Vulcan—one-time hypothetical planet orbiting the Sun closer than does Mercury ⚄ named for the Roman god of fire

W

Wane (see Phase)

Water cycle—continuous movement of water within the atmosphere, surface, and interior of a planet

Wavelength—distance between sequential crests or troughs in a wave

Wax (see Phase)

Weak nuclear force—causation at a distance significant only on a subatomic scale; responsible for radioactive decay

White dwarf—stellar end state in which a low-mass star has collapsed into the smallest possible sphere of normal matter; nuclear energy is no longer produced, but the star glows hot due to residual heat

Widmanstätten pattern—large crystalline structure within a meteorite indicative of slow solidification within liquid metal in interplanetary space

X

X-ray (see Electromagnetic spectrum)

Y

Year Star—Chinese calendrical use of the planet Jupiter owing to the fact that the planet travels through one Chinese constellation per year over a 12-year cycle

Z

Zodiacal light—sunlight reflected from interplanetary dust in the ecliptic plane; most easily seen in the western sky after sunset, or eastern sky before sunrise, as an elongated cone extending up from the horizon

Zone—high-albedo cloud band encircling a Jovian Planet parallel to its equator

Illustration Credits

Unless otherwise described in the caption, images are digital, photographic positives. They are made in visible light unless indicated. However, image processing may enhance color and contrast in order to better show subtle features. When possible, false-color is avoided. Images are ground-based unless a space mission is referenced in the caption. Parentheses below mean that an original illustration is modified for use in this book.

ABM = Australian Bureau of Meteorology
AGU = American Geophysical Union
ARC = Ames Re search Center
BGS = British Geologic Survey
CAS = Chinese Academy of Sciences
ESA = European Space Agency
ESO = European Southern Observatory
GSFC = Goddard Space Flight Center
IAU = International Astronomical Union
JAEA = Japanese Atomic Energy Agency
JAXA = Japanese Aerospace Exploration Agency
JHU-APL = Johns Hopkin University-Applied Physics Laboratory
JPL-Caltech = Jet Propulsion Laboratory-California Institute of Technology
JSFC = Johnson Space Center
KSFC = Kennedy Space Flight Center
LBNL = Lawrence Berkeley National Laboratory
NAIC = National Astronomy and Ionosphere Center
NAOJ = National Astronomical Observatory of Japan
NAS = National Academy of Sciences
NASA = National Aeronautics and Space Administration
NIH = National Institute of Health
NOAA = National Oceanic and Atmospheric Administration
NOAO = National Optical Astronomical Observatory
NRAO = National Radio Astronomy Observatory
NSO = National Solar Observatory
RAS = Russian Academy of Sciences
STScI = Space Telescope Science Institute
SwRI = Southwest Research Institute
USDE = United States Department of Energy
USDI = United States Department of the Interior
USGS = United States Geological Survey
WMAP = Wilkinson Microwave Anisotropy Probe Science Team

Introduction Declan Deval
1-1 NASA/WMAP
1-2 NASA
1-3 NASA/ESA
1-4 NASA/STScI
1-5 Arthur Mouratidis*

1-6 NASA/JPL-Caltech
1-7 ESA
1-8 NASA/HPL-Caltech
1-9 ESO
1-10 Gomes, R.; Levison, H. F.; Tsiganis, K.; and Morbidelli, A. "Origin of the Cataclysmic Late Heavy Bombardment Period of the Terrestrial Planets." *Nature*. **435**, 7041 (2005).

2-1 Stellarium software
(authors)
2-2 William Spurr
2-3 William Spurr
2-4 anonymous (16th century). District Museum, Toruń, Italy.
2-5 William Spurr
2-6 anonymous (1610). GL Archive.
2-7 William Spurr
2-8 authors
2-9 William Spurr
2-10 Godfrey Kneller (1646-1723). National Portrait Gallery, London, UK.
2-11 frontispiece: Holden, Edward S. *Sir William Herschel: His Life and Works*. New York, NY: Charles Scribner's Sons (1881).
2-12 ESA
2-13 NASA
2-14 Science Photo Library
2-15 Science Photo Library
2-16 BGS
2-17 NASA/Peter Reid

3-1 annonymous (mid 100 s). Museo Nazionale, Rome, Italy.
3-2 Justus Sustermans (1597-1681). Uffizi Gallery, Florence, Italy.
3-3 Pxhere
(authors)
3-4 AllrightImages
3-5 Isaac Newton (1686). Science Museum, UK.
3-6 Theodore Goetz (1823). Florilegius.
3-7 Mya Dicus
3-8 Mya Dicus
3-9 Mya Dicus
3-10 Mya Dicus
3-11 NASA/Joel Kowsky
3-12 frantic 19
3-13 NASA
3-14 Johan Swanepoel

4-1 NASA/JSFC
4-2 NASA
4-3 Anthony Ayiomamitis
4-4 NASA/GSFC
4-5 Florence McGinn
4-6 NAOJ
4-7 NAS
4-8 ESA

4-9 NASA
4-10 Alexandru Barbovschi
4-11 NASA/GSFC
4-12 NASA/GSFC
4-13 NASA/JHU-APL
4-14 NASA/JSFC
4-15 NASA/GSFC
4-16 NASA
4-17 NASA
4-18 NASA
4-19 NASA
4-20 NASA/JSFC
4-21 NASA
4-22 NASA/JPL-Caltech
4-23 NASA
4-24 NASA
4-25 CAS/Science and Application Center for Moon and Deep Space
Exploration

5-1 Mlungisi Louw/Netwerk24
5-2 G. Hudepohl/ESO
5-3 LBNL
5-4 NASA/JHU-APL
5-5 NASA/JHU-APL
5-6 ESA
5-7 NASA/JHU-APL
5-8 NASA/JPL-Caltech
5-9 NASA/JHU-APL
5-10 NASA

6-1 Juan Carlos Casado/The World at Night
6-2 Galileo Galilei (1623). Museo Galileo, Florence, Italy.
6-3 NASA
6-4 NSO
6-5 NASA
(authors)
6-6 Burroughs, E. R. *Lost on Venus*. New York, NY: Ace Books (1935). [original publication date]
6-7 NASA/JPL-Caltech
6-8 NASA
6-9 A Loose Necktie*
6-10 NASA/ARC
6-11 ESA
6-12 NASA/JPL-Caltech
6-13 NASA/JPL-Caltech
6-14 NASA/JPL-Caltech
6-15 JAEA

7-1 NASA/JPL-Caltech
7-2 iStock
7-3 NASA

7-4 USGS
7-5 USGS
7-6 USGS
7-7 NOAA
7-8 NASA
7-9 NASA
7-10 NASA
7-11 Rober Nufer
7-12 NOAA
7-13 NOAA
7-14 ABM
(authors)
7-15 ABM
(authors)
7-16 NOAA
7-17 NASA
7-18 AGU
7-19 Fritz Geller-Grimm*

8-1 NASA/Calvin J. Hamilton
8-2 NASA/JPL-Caltech
8-3 NASA/JPL-Caltech
8-4 NASA/JPL-Caltech
8-5 NASA/JPL-Caltech
8-6 NASA/JPL-Caltech
8-7 NASA/JPL-Caltech
8-8 ESA
8-9 NASA/JPL-Caltech
8-10 NASA/JPL-Caltech
8-11 NASA/JPL-Caltech
8-12 NASA/JPL-Caltech
8-13 NASA/JPL-Caltech
8-14 NASA/JPL-Caltech
8-15 NASA/JPL-Caltech
8-16 NASA/JPL-Caltech

9-1 Davida Hardy
9-2 ESO
9-3 USDE
9-4 NASA
9-5 NASA
9-6 NASA
9-7 USGS
9-8 NASA
9-9 ESA
9-10 Sputnik Images*
9-11 JAXA
9-12 NASA
9-13 NASA
9-14 NASA
9-15 NASA/JPL-Caltech

9-16 NASA/JPL-Caltech
9-17 JAXA
9-18 NASA/JPL-Caltech
9-19 JAXA
9-20 ESA
9-21 NASA/JPL-Caltech/David Fuchs
9-22 USDI
9-23 Tomruen*

10-1 NASA
10-2 NASA/JPL-Caltech
10-3 NASA/JPL-Caltech
10-4 William Spurr
10-5 NASA
10-6 NASA
10-7 NASA/JPL-Caltech
10-8 NASA/JPL-Caltech
10-9 NASA
10-10 NASA/JPL-Caltech
10-11 William Spurr
10-12 NASA/JPL-Caltech
10-13 NASA/JPL-Caltech
10-14 NASA/JPL-Caltech
10-15 NASA/JPL-Caltech
10-16 NAOJ
10-17 ESA/NASA
10-18 NASA
10-19 NASA/JPL-Caltech
10-20 Galileo Galilei (1626). NASA/GSFC
10-21 NASA/STScI
10-22 NASA/JPL-Caltech
10-23 NASA/JPL-Caltech
10-24 NASA/JPL-Caltech
10-25 NASA/JPL-Caltech
10-26 NASA/JPL-Caltech
10-27 NASA/JPL-Caltech

11-1 NASA/JPL-Caltech
11-2 NASA/JPL-Caltech
11-3 NASA/JPL-Caltech
11-4 NASA/JPL-Caltech
11-5 NASA/JPL-Caltech
11-6 NASA/JPL-Caltech
11-7 NASA/JPL-Caltech
11-8 NASA/JPL-Caltech
11-9 NASA/JPL-Caltech

12-1 NASA
JAXA
(Kevin M. Gill*)
12-2 NASA/JPL-Caltech

12-3 NASA/JPL-Caltech
12-4 The Singing Badger*
12-5 NASA/ESA
12-6 Philipp Salzgeber*
12-7 NASA/JPL-Caltech
12-8 NASA/JPL-Caltech
12-9 NASA/JPL-Caltech
12-10 NASA/JPL-Caltech
12-11 NASA/JPL-Caltech
12-12 ESA/NASA
12-13 NASA/JPL-Caltech
(authors)
12-14 NASA/JPL-Caltech
12-15 NASA/JPL-Caltech
12-16 NASA/JPL-Caltech
12-17 NASA/JPL-Caltech
12-18 NASA/JPL-Caltech
12-19 NASA/JPL-Caltech
12-20 NASA/JPL-Caltech

13-1 NASA
13-2 Bayeux Tapestry (1066). Musée Bayeaux, Bayeaux, France.
13-3 John Faber (1722)
13-4 Alpha Stock
13-5 Christopher*
13-6 Andrzej Mirecki*
13-7 NASA/JPL-Caltech
13-8 ESO/M. Kornmesser
13-9 ESA/NASA
13-10 NASA/JPL-Caltech
(authors)
13-11 NASA/NOAO
13-12 NASA
ESO
(authors)
13-13 NASA/JPL-Caltech
13-14 NASA/JPL-Caltech
13-15 NASA/ESA
NASA   .
(authors)
13-16 Iván Éder*
13-17 ESA
13-18 NASA
13-19 Planetary Society
RAS/Ted Stryk
NASA/JPL-Caltech
ESA/Emily Lakdawalla
NASA/JPL-Caltech
NASA/JPL-Caltech
NASA/JPL-Caltech
(Emily Lakdawalla)

13-20 NASA/STScI
13-21 NASA
NASA/STScI
(authors)

14-1 Encrenaz, T. *Annual Review of Astronomy and Astrophysics*. **46**, 57 (2008).
14-2 NASA
14-3 NASA
14-4 NASA/JHU-APL
14-5 NASA
14-6 NASA/STScI
14-7 NASA/JHU-APL
14-8 NASA/JHU-APL
14-9 NASA/JHU-APL
14-10 ESO
14-11 NASA/ESA
14-12 NASA
14-13 NASA/JPL-Caltech

15-1 Anonymous (8th century). County Museum of Art, Los Angeles, California, USA
15-2 Yuliana Ivakh
15-3 NASA
(Dave Jarvis)
15-4 Needpix.com
15-5 Needpix.com
15-6 "The Solar Eclipse of May 29, 1919, and the Einstein Effect," *Scientific Monthly*. **10**, 4 (1920).
15-7 tOrange.biz*
15-8 Ildar Sagdejev*
15-9 D. D. Shaihulud*
15-10 ESO
15-11 W. Carter*
15-12 NASA/GSFC
15-13 NASA
15-14 Luc Viatour*
15-15 ESA
15-16 NASA/GSFC
15-17 NASA/ESA
15-18 NASA/GSFC

16-1 Marie-Lan Nguyen/Metropolitan Museum of Art*
16-2 Planetkid*
16-3 ESA
16-4 ESO
16-5 NASA/JPL-Caltech
16-6 ESO
16-7 NASA/ARC
16-8 NASA/JPL-Caltech
16-9 NASA/GSFC
16-10 NASA/JPL-Caltech
16-11 NASA

16-12 NASA/JPL-Caltech
16-13 NASA/JPL-Caltech
16-14 ESO
16-15 NASA
16-16 NASA/JPL-Caltech
16-17 ESO
16-18 NASA/JPL-Caltech
16-19 NASA/ESA
16-20 Universito de Puerto Rico/NAIC*

17-1 NASA
17-2 USDI
17-3 Schokraie, E.; Warnken, U.; Hotz-Wagenblatt, A.; Grohme, M. A.; Hengherr, S., *et al.* "Comparative Proteome Analysis of Milnesium Tardigradum in Early Embryonic State Versus Adults in Active and Anhrobiotic State." *PLoS ONE.* **7**, 9 (2012).*
17-4 Woudloper*
17-5 University of Rochester
17-6 Yuliana Ivakh
17-7 NASA/GSFC
17-8 SwRI
17-9 ESO
17-10 NASA/JPL-Caltech
17-11 NASA/JPL-Caltech
17-12 Universito de Puerto Rico/NAIC*
17-13 NASA/JPL-Caltech
17-14 NASA/JPL-Caltech

A-1 ESA/Annekele Floc'h
A-2 NASA/KSC
A-3 CNSA
A-4 NASA
A-5 JAXA
A-6 NASA/Chris Gunn
A-7 NASA/JPL-Caltech
A-8 NASA
A-9 NASA/JPL-Caltech
A-10 NASA
A-11 NASA
A-12 ESA
A-13 NASA
A-14 Emerezhko*
A-15 NASA/JPL-Caltech

*https://creativecommons.org/licenses/by-sa/4.0/

Index

absorption, 79, 89, 91, 93
acceleration, 29–32, 34–5, 38
accretion, 7–8, 70
acid, 78–79, 243–4
acidophile, 244
Active Asteroid, 196
Adam, 1
Adams, John, 155
Adity-L1, 253
Aeolis Mons, 109
aerial, 119
aerosol, 143
Africa, 87, 120, 189, 227
aggregation, 7, 207
airliner, 91
Akatsuki, 80
albedo, 66, 141, 160, 170, 174, 202, 204–5
Alfvén, Hannes, 189
algae, 245
Alighieri, Dante, 236
alkaliphile, 244
Allan Hills, 243–4
Allen, James, 91, 93, 140
Allen, Paul, 255
Alps, 219
altitude, 69, 89, 91, 143–4
Amalthea, 125, 172
Amino acid, 243
ammonia, 141, 143, 174
Amtor, 76
anaerobe, 244
Andromeda Galaxy, 1, 10, 232
angular measurement, 22, 35–6, 184, 211
anisotropy, 2
annihilation, 213, 216
annular eclipse, 50–1
Ansari, Anousheh, 255
Ansari XPRIZE, 255
Antarctica, 71, 118, 243–4, 249
Anthropocene, 10
antimatter, 216
anti-tail, 191
aphelion, 16, 63–4, 187–8, 202
Aphrodite, 73
apogee, 45, 47, 50
Apollo asteroid, 121
Apollo program, xvii, 44, 85, 123
aquifer, 248

archaea, 245
Archean, 10
Arecibo Observatory, 63, 71, 78
Ares, 101, 104, 109
argon, 218
Ariel, 176
Aristotle, 29–30
Arizona, 107, 119
Arrokoth, 205–6
arroyo, 109
Artemis Accords, 248–50
ash, 213
Asimov, Isaac, 160, 163
Assayer, 74
asteroid, xviii, 19–20, 29, 32, 37
Asteroid Belt, 19–20, 37–8
Asteroid Terrestrial-impact Last Alert System, 187
Astraea, 19
astrobiology, 183, 243, 246–8
astrochemistry, 189
astrology, 189
astrometry, 227, 229
astronaut, 38, 64, 104, 107, 110, 139, 232
astronomer, xviii–xix, 1, 10, 13, 17, 19–20, 29, 38, 66,
    74–5, 101, 104, 124, 155, 160, 163, 166, 181,
    188–9, 199, 201, 206, 211–12, 227, 244, 247
Aten asteroid, 121, 123
Atlantic Ocean, 87
Atlas, 178
atlas, 10
atmosphere, xviii, 8, 10, 49, 65, 69, 74–5, 78–81, 89–91,
    93, 104–7, 110, 115, 117–19, 121, 135, 138–4,
    155–8, 160–1, 163, 165, 173–4, 181, 201–2,
    204, 214, 219, 231–2, 234, 236, 243, 245–6,
    248–51
atom, 2, 69, 79, 89, 214–16, 188, 211, 218
Audhumla, 1
aurora, 93–4, 140–1, 158, 171
Austria, 76
Automatic Planet Finder, 232
Axion, 255
axis, 4, 91, 93, 138, 140, 155, 158, 160, 163

Babylonia, 63, 73, 101, 135
band, 138–9, 141
barge, 143
Barnard's Star, 229
Barringer Crater, 119–20

Printed in the United States
by Baker & Taylor Publisher Services

Printed in the United States
by Baker & Taylor Publisher Services